Nitrate Contamination

Exposure, Consequence, and Control

NATO ASI Series

Advanced Science Institutes Series

A series presenting the results of activities sponsored by the NATO Science Committee, which aims at the dissemination of advanced scientific and technological knowledge, with a view to strengthening links between scientific communities.

The Series is published by an international board of publishers in conjunction with the NATO Scientific Affairs Division

A Life Sciences B Physics	Plenum Publishing Corporation London and New York

C Mathematical and
   Physical Sciences
D Behavioural and
   Social Sciences
E Applied Sciences
	Kluwer Academic Publishers
	Dordrecht, Boston and London

F Computer and
   Systems Sciences
G Ecological Sciences
H Cell Biology
I Global Environmental
   Change
	Springer-Verlag
	Berlin Heidelberg New York
	London Paris Tokyo Hong Kong
	Barcelona Budapest

NATO-PCO DATABASE

The electronic index to the NATO ASI Series provides full bibliographical references (with keywords and/or abstracts) to more than 30 000 contributions from international scientists published in all sections of the NATO ASI Series. Access to the NATO-PCO DATABASE compiled by the NATO Publication Coordination Office is possible in two ways:

– via online FILE 128 (NATO-PCO DATABASE) hosted by ESRIN,
  Via Galileo Galilei, I-00044 Frascati, Italy.

– via CD-ROM "NATO-PCO DATABASE" with user-friendly retrieval software
  in English, French and German (© WTV GmbH and DATAWARE Technologies
  Inc. 1989).

The CD-ROM can be ordered through any member of the Board of Publishers or through NATO-PCO, Overijse, Belgium.

Series G: Ecological Sciences Vol. 30

Nitrate Contamination

Exposure, Consequence, and Control

Edited by

Istvan Bogárdi

Department of Civil Engineering
University of Nebraska-Lincoln
Lincoln, Nebraska 68588-0531
USA

Robert D. Kuzelka

University of Nebraska Water Center
Lincoln, Nebraska 68583-0844
USA

Technical Editor:

Wilma G. Ennenga

Department of Civil Engineering
University of Nebraska-Lincoln
Lincoln, Nebraska 68588-0531
USA

Springer-Verlag
Berlin Heidelberg New York London Paris Tokyo
Hong Kong Barcelona Budapest
Published in cooperation with NATO Scientific Affairs Division

Proceedings of the NATO Advanced Research Workshop on Nitrate Contamination: Exposure, Consequences, and Control held at Lincoln, Nebraska (USA) from September 9–14, 1990.

ISBN 3-540-53088-6 Springer-Verlag Berlin Heidelberg New York
ISBN 0-387-53088-6 Springer-Verlag New York Berlin Heidelberg

© Springer-Verlag Berlin Heidelberg 1991
Printed in Germany

Typesetting: Camera-ready by authors
31/3140-543210 – Printed on acid-free paper

CONTENTS

I. OVERVIEW OF PROBLEM

II. EXPOSURE ASSESSMENT

III. HEALTH CONSEQUENCES

LIST OF ABBREVIATIONS

BOD	biological oxygen demand
CAG	chronic atrophic gastritis
COD	carbon oxygen demand
DEA	denitrifying enzyme activity
DOC	dissolved organic carbon
ER	electrical resistivity
FEM	frequency domain electromagnetics
MCL	maximum contaminant level
NA	N-nitrosamines
NDELA	N-nitrosodiethanolamine
NHL	non-Hodgkin's Lymphoma
NMAP	N-nitrosomethylamino propionitrile
NNC	N-nitroso compounds (Mull)
NOC	N-nitroso compounds (Crespi)
NPC	nasopharingeal carcinoma
NPRO	N-nitroso proline
RfD	reference dose
TEM	time domain electromagnetics
TKN	total kjeldahl nitrogen
TOC	total organic carbon
TOD	total oxygen demand
TSS	total suspended solids
WSC	water soluble carbon

PREFACE

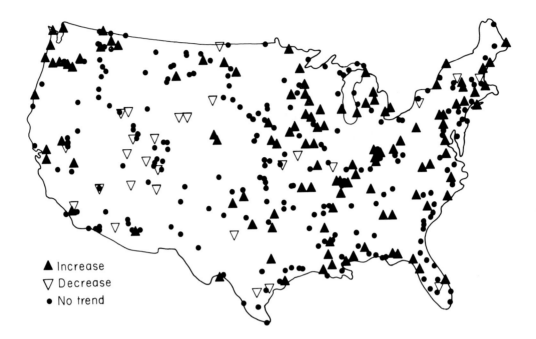

Fig. 3. Trends in flow-adjusted nitrate concentrations from 1974 to 1981 (Smith et al., 1987)

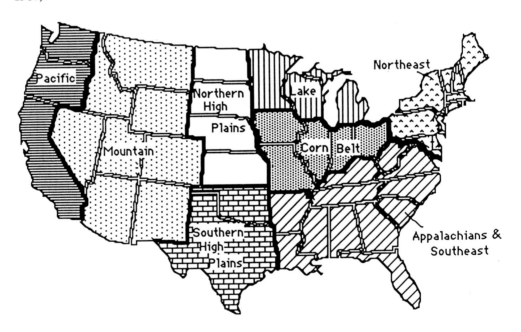

Fig. 4. Geographical regions of the contiguous United States

2 Case Studies

2.1 The Northeastern States (Fig. 5)

Approximately one-fifth of the nation's population lives in the Northeast, the most urban area in the United States. The population is concentrated in an area between Boston, Massachusetts and Washington, D.C., where heavy and light manufacturing and high technology and service industries prevail. Beyond the megalopolis are forests, which cover about 60% of the Northeast's 47 million hectares (U.S. Department of Agriculture (USDA), 1988), and farm-land. Dairy farming is a major agricultural activity and a large proportion of the 6.3 million hectares of cropland are cropped to hay (U.S. Department of Commerce, 1989). Corn, the crop with the highest nitrogen fertilizer requirement, was grown on less than one million hectares in 1987 (U.S. Department of Commerce, 1989).

Fig. 5. Locations of cited investigations in the Northeastern States

Because the rolling, rocky land with its thin soil and the Appalachian Mountains make most of New England unsuitable for mechanized agriculture, it is localized in areas like the Connecticut River Valley. In an investigation of the impact of fertilizer leachates from both urban and farmland sources on ground-water quality in the Connecticut River Valley, DeRoo (1980) found that nitrate-N levels in ground water downgradient from farmland treated with excess N-fertilizer ranged from 10 to more than 20 mg/L.

On Long Island, New York, both agricultural areas and wild lands have been developed for residential areas. Ragone et al. (1976) determined that nitrate levels were increasing in the ground water beneath both the sewered and unsewered areas of Suffolk County. The majority of the contamination in the unsewered agricultural areas was derived from N fertilizers, although animal-derived N, presumably from septic-tank leachates, was identified in several wells (Kreitler et al., 1978). Flipse et al. (1984) estimated that in a sewered residential area of Suffolk County at least 60% of the N-fertilizer applied to turfgrass leached. Nitrogen isotopes confirmed that agronomic (fertilizer and mineralized N) leachates were the major source of the nitrate and that the impact of wastes from high densities of domestic animals and sanitary sewer lines on ground-water quality was minor.

Both nitrate and herbicide detections have been reported in ground water in watersheds of the Susquehanna River Basin in Lancaster County, Pennsylvania (Fishel and Lietman, 1986). Nitrate concentrations were significantly higher in the agricultural areas than in the nonagricultural areas. Wells penetrating carbonate rock generally contained three times more nitrate-N than wells penetrating noncarbonate rock. The sources of nitrate contamination were considered but not confirmed; however, it is known that large amounts of manure were applied to these row-cropped fields and pastures.

Ritter and Chirnside (1984) measured nitrate in water-table wells in two of Delaware's three counties. Sampling was concentrated in forested, cropped (corn and soybeans), intensive broiler (poultry) production, and unsewered residential areas. The highest nitrate concentrations occurred in areas with either intensive broiler production or intensive crop production. In four of the five identified ground-water problem areas, leachates from poultry manure appeared to be the major contributor of nitrate to the ground water. Leachates from commercial fertilizer and septic tanks also contributed to the contamination. Ancillary chloride and $\delta^{15}N$ analyses were used to differentiate agronomic (fertilizer and mineralized nitrate) and animal waste sources. Ritter and Chirnside (1984) suspected nitrate contamination occurs in other vulnerable areas of the Delmarva Peninsula.

2.2 The Appalachian and Southeastern States (Fig. 6)

This 12-state area comprises approximately 142 million hectares. More than half (57%) of the region is forested, about 10% is pasture and range land, and about 20% is cropped (USDA, 1988). In 1987 the 6.5 million hectare harvest of soybeans and cotton represented 23% and 34%, respectively, of the nation's production of these commodities (U.S. Department of Commerce, 1989). Corn was grown on 1.9 million hectares. Almost three-quarters of the nation's broiler chickens are raised in the region, which is undergoing a transformation from an agricultural to industrial economy.

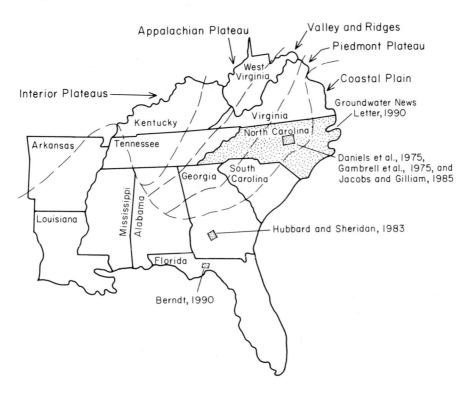

Fig. 6. Locations of cited investigations in the Appalachian and Southeastern States

The farmland is concentrated in the Coastal Plain and the Piedmont Plateau (Fig. 6). Despite the sandy soils, which require heavy applications of fertilizer, the high amounts of rainfall (1270 mm/year), and the rapid development of irrigation, there are few incidences of groundwater nitrate contamination (Hubbard and Sheridan, 1989). Fortunately, much of the region is underlain by shallow retarding layers which perch the recharge. Lateral flows divert the recharge to swamps where vegetation utilizes the nutrients. Because the aquifers throughout much of the southeast are confined or semiconfined, only very shallow or perched ground water appears affected by nitrate contamination. Significant vertical stratification and lateral concentration gradients of nitrate in shallow subsurface waters beneath fertilized plots on the Tifton Upland in Georgia confirmed the hypothesis. The loss in nitrate was ascribed partially to denitrification and uptake by deep-rooted vegetation.

In North Carolina, Daniels et al. (1975), Gambrell et al. (1975), and Jacobs and Gilliam (1985) found that substantial nitrate was leached below the root zone to subsurface flows where most of the nitrate was depleted via denitrification. The preliminary results of a statewide ground-water monitoring program supported these findings. To date nitrate levels in only 3.3% of the more than 3,300 wells analyzed in 11 counties exceeded the MCL (The Groundwater Newsletter, 1990). Eventually nitrate will be measured in 8,000 private wells in 29 North Carolina counties. Jacobs and Gilliam (1985) found that despite a 400% increase in fertilizer use in North Carolina since 1945, nitrate levels in streams did not increase.

Berndt (1990) found that nitrate concentrations increased in both the surficial aquifer and Upper Floridan aquifers beneath a sprayfield treated with secondary sewage effluent and commercial fertilizer. Nitrate concentrations in the surficial aquifer exceeded 10 mg/L NO_3-N and were an order of magnitude higher than in the surficial aquifer outside the irrigated site. Concentrations were lower in the shallow wells in the Upper Floridan aquifer but still significantly higher than in wells at similar depths located outside the irrigated site. There was little difference in the nitrate concentrations in wells in the Upper Floridan aquifer with depths of more than 30 meters whether located inside or outside the irrigated site. Berndt (1990) compared the nitrogen isotopic ratios of the nitrate in ground water beneath the southeast sprayfield to those beneath a second sprayfield receiving only the secondary sewage effluent in order to determine the relative proportions of nitrate contributed by the two N sources. The significantly heavier ratios in the ground water beneath the effluent-fertilized, southwest sprayfield were in the range characteristic of animal waste (Fig. 7). The lower values beneath the southeast sprayfield treated with both effluent and commercial fertilizer were between the ranges for animal waste and commercial fertilizer sources and showed an identifiable contribution from inorganic fertilizer.

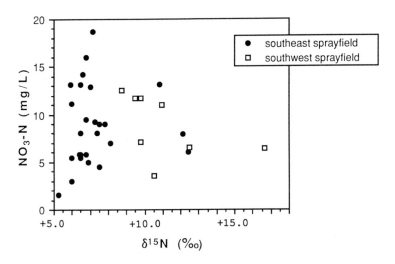

Fig. 7. Nitrate concentrations and isotope ratios in ground-water samples beneath two sprayfields receiving different fertilizer treatments (Berndt, 1990)

2.3 The Corn Belt States (Fig. 8)

Level land, abundant rainfall, fertile soils, and hot, humid summers combine to make this some of the most productive cropland in the world. With 58% of the approximately 67.3 million hectares in cropland, not only does the region have the largest proportion of cropland in the United States, but it also is the only region where corn commands the largest proportion (30%) of the cropland (U.S. Department of Commerce, 1989). The almost 12 million hectares planted to corn in 1987 represented half of the nation's corn acreage and 54% of its harvest. Soybeans were grown on about the same number of acres and that harvest represented about 60% of the U.S. crop. The area also produced more than half of the hogs sold in the United States (U.S. Department of Commerce, 1989).

Lee and Nielsen's (1989) assessment indicated that ground water in large areas of four of the five states was very vulnerable to contamination from agricultural sources of nitrate. Nitrate contamination in Ohio has been well-documented by Baker et al. (1989), who measured nitrate concentrations in 16,166 private wells to thoroughly assess the extent of nitrate contamination in Ohio's ground water. Only 2.7% of the 14,478 drinking-water wells exceeded the MCL for nitrate-N. Concentrations in 10% of the wells ranged from 3 to 10 mg/L. Baker and his collaborators reported several interesting observations: the lack of an association between land use and ground-water nitrate levels in northwestern Ohio where row-crop agriculture is most intensive; higher nitrate levels in wells completed in unconsolidated alluvial aquifer systems, in older wells, and in wells with completion depths of less than 15 meters; higher nitrate levels in wells more than 60 meters from septic tanks than in wells within 60 meters of septic tanks; and a higher frequency of nitrate-N levels above 10.0 mg/L in wells in close proximity to cattle feeding areas. Their data supported the conclusions of Gillham and Webber (1969), Kreitler (1975), and Exner et al. (1985) that shallow alluvial wells in river and stream valleys and poorly constructed wells sited near animal corrals are highly vulnerable to point sources of contamination.

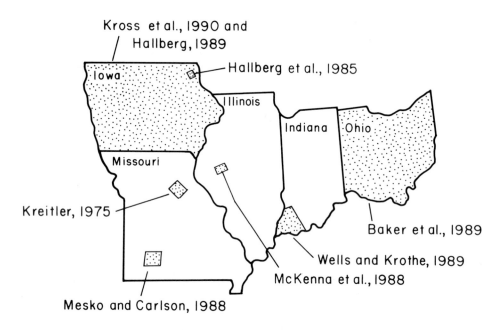

Fig. 8. Locations of cited investigations in the Corn Belt States

Power and Schepers (1989) attributed the relative lack of agrichemical contamination in ground water of the eastern Corn Belt to the widespread use of tile drains, which intercept nitrate-contaminated recharge and discharge it to surface waters. Logan et al. (1980) reported that nitrate levels in tile drain discharge increased with increased N-fertilizer application rates. Losses of 20 to 100 kg NO_3-N/ha/yr were common beneath tile-drained fields at numerous sites in several north-central states.

In Indiana and Illinois, investigations of ground-water nitrate contamination have focused on relatively small areas. The contamination probably is related more to the hydrogeology of the site and availability of sources than to statewide occurrence. Bicki et al. (1984) showed that home sewage disposal in densely populated subdivisions in shallow ground-water karst regions in Florida effected ground and surface-water quality. In the karst area of southern Indiana, where heavy rains induce rapid recharge through the macropores in the karst as evidenced by muddy ground water, nitrogen isotope ratios were employed to investigate the potential sources of nitrate contamination (Wells and Krothe, 1989). Isotope ratios were measured in the fall and again the following spring. While the nitrogen isotope ratios of the May samples suggested there were inputs of fertilizer-derived N after spring fertilization, the heavier isotopic values obtained the previous September lead to the conclusion that septic field leachates were the most probable contributors of nitrate in this area of pronounced preferential flow. Another plausible mechanism for the lower nitrate concentrations and enrichment in N-15 in September could be denitrification of agronomically-derived nitrogen in the shallow ground water of the area.

McKenna et al. (1988) assessed the occurrence of agrichemicals in ground water beneath irrigated and fertilized fields in a hydrogeologically sensitive area of west-central Illinois. Nitrate levels were considerably higher in shallow wells downgradient from irrigated, fertilized corn fields than in shallow wells adjacent to nonirrigated, fertilized corn fields. The effects of fertilizer leachates on ground-water quality were obvious in a comparison of mean nitrate levels in control wells sited in a natural area and a monitoring well downgradient from a fertilized corn field. While concentrations averaged 0.5 mg/L NO_3-N in the control wells, they ranged from 11 to 22 mg/L NO_3-N in the downgradient well. Rotating corn to soybeans significantly lowered the levels of nitrate in the ground water although the levels still remained above the MCL. This observation suggested that leaching of nitrate continued through the next growing season.

Although the nitrate contamination in the Big Spring karst area of northeastern Iowa has received much visibility, there is significant nitrate contamination in both karst and nonkarst regions of Iowa. The State-Wide Rural Well-Water Survey, a one-time representative sampling of 684 domestic wells in Iowa's 99 counties, showed for the first time the extent of nitrate contamination (Kross et al., 1990). Nitrate concentrations exceeded the MCL in approximately 18% of the sampled wells. The highest incidence of contamination (38%) occurred in northern Iowa while the lowest incidence (6%) was in north-central Iowa. Wells less than 15 meters deep had a higher incidence of contamination (35%) than did deeper wells. Nearly half of the sampled wells contained coliform bacteria.

Hallberg (1989) also showed that nitrate contamination occurs primarily in the uppermost 15 meters of the aquifer. He ascribed the vertical distribution of nitrate to the hydrogeology of both the unsaturated zone and aquifer and to the potential for increased denitrification with depth in bedrock aquifers. Hallberg and Hoyer (1982) also documented significant nitrate contamination in aquifers overlain by less than 15 meters of glacial deposits.

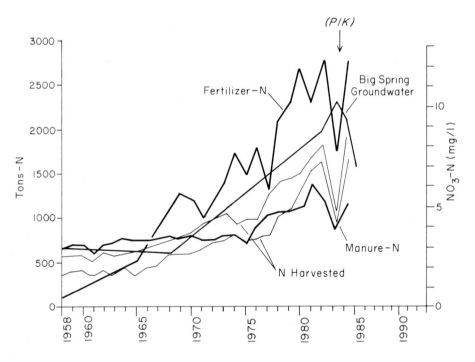

Fig. 9. Fertilizer use in the Big Spring Basin and average annual nitrate concentration in ground water at Big Spring, Iowa (Hallberg, 1989)

Hallberg et al. (1985) attributed the increases in nitrate concentrations in the ground water of the Big Spring area from the late-1960s to mid-1980s (Fig. 9) to increased fertilizer usage. Unlike suspended solids and pesticides, peak nitrate inputs occurred after the run-in event when the hydrograph was in recession and infiltration recharge dominated. They concluded

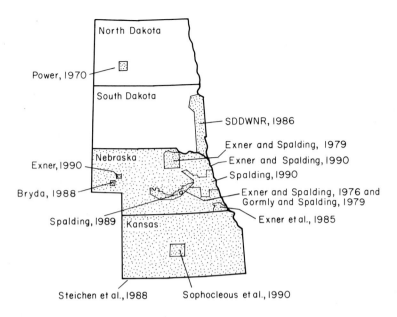

Fig. 11. Locations of cited investigations in the northern Plains States

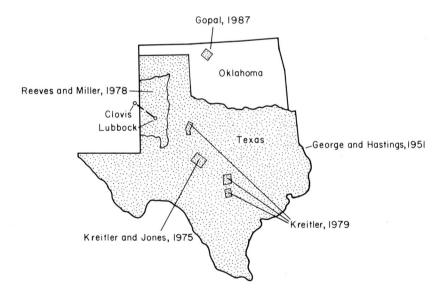

Fig. 12. Locations of cited investigations in the southern Plains States

exceeded the MCL. Because sampling sites were not randomly selected across the state, and the density of samples was greater in areas of known or suspected nitrate contamination, the data are not representative. Slightly more than half of the elevated nitrate concentrations occurred in areas highly vulnerable to nonpoint nitrate contamination (Fig. 14). These areas are characterized by irrigated corn monoculture on well to excessively well-drained soils and a vadose zone less than 15 meters thick.

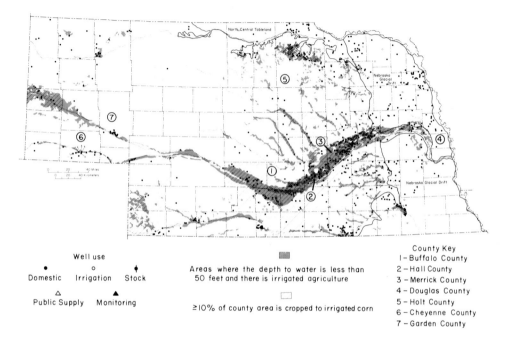

Fig. 14. Occurrence of nitrate-nitrogen concentrations greater than 10 mg/L NO$_3$-N in Nebraska's ground water (Exner and Spalding, 1990)

The largest expanse of nitrate-contaminated ground water, more than 202,000 contiguous hectares, occurs within the area of high vulnerability on the north side of the Platte River in the central Platte Valley (Fig. 14). An initial sampling of about 400 wells in the central Platte valley in 1974 (Exner and Spalding, 1976) and the resampling of the same wells a decade later (Exner, 1985) documented increases in nitrate concentrations and in the areal extent of the contamination. Within the contaminated area, average nitrate-N concentrations increased 0.4 to 1.0 mg/L NO$_3$-N/yr to average concentrations that were 1.5 to more than 2 times higher than the MCL. The extent of the contamination more than doubled. While nitrate concentrations beneath fields in the Hall County Water Quality Project appeared to have stabilized in 1983 at the conclusion of the five-year management study (Bockstadter et al., 1984), there still was an overall increase in nitrate concentrations within the project area between 1974 and 1984.

Using the natural variations in nitrogen isotopes, Gormly and Spalding (1979) concluded that commercial fertilizer was the primary cause of the contamination in the high nitrate zones of Buffalo, Hall, and Merrick Counties (Fig. 14). They noted that isotopic fractionation was spatially associated with changes in soil texture. In the centers of the contaminated zones where nitrate concentrations were highest, the isotopic ratios were lowest being just slightly heavier than the signature of commercial anhydrous ammonia fertilizer. These areas were characterized by excessively well-drained soils, which promote rapid infiltration with only limited isotopic fractionation. Moving away from the centers of the zones, the soils became less well-drained, the nitrate concentrations were lower, and the isotopic ratios increased. This trend of decreasing concentrations and increasing isotopic ratios is characteristic of denitrification. Significant negative correlations were obtained between $\delta^{15}N$ values and nitrate concentrations and between $\delta^{15}N$ values and depth to water. These associations support the contention that denitrification can be a major mechanism for nitrogen loss in irrigated agriculture where fine-textured soils have claypan horizons or fine-textured soils overlie a shallow (<1.5 m) water table.

A recent investigation in the Lower Platte Valley (Spalding, 1990) substantiated the occurrence of denitrification in very shallow water-table areas and revealed a potential pitfall of reliance on generalized methods to predict pollution potential. The area between the Platte and Elkhorn Rivers in Douglas County (shaded area in Fig. 14) is cropped to corn and irrigated with ground water that lies less than 2 meters below the surface. With a DRASTIC (Aller et al., 1985) Index above 160 and much of the area exceeding 180, the ground water can be characterized as having a high potential for contamination. The presence of trace levels of atrazine in the ten irrigation wells sampled for atrazine indicated that conditions are favorable for the leaching of agrichemicals. Nitrate concentrations in the majority of the 15 irrigation wells sampled, however, were less than 1 mg/L NO_3-N and were between 1 and 5 mg/L in a few wells. Only three wells had concentrations above 7 mg/L NO_3-N. Their highly fractionated isotopic ratios ($+15$ to $+32$‰) suggest a high level of denitrification. While denitrification in these water-logged, irrigated soils controls the concentrations of nitrate leaching to the ground water, there are some indications that the aquifer is slightly reducing and denitrification also occurs in the ground water.

Along the northern edge of a high nitrate (>10 mg/L NO_3-N) zone in the Central Platte Valley, Gormly and Spalding (1979) reported a sharp decrease in nitrate concentrations. *In situ* oxidation potential measurements indicated that within the regional flow field there is a an abrupt change in nitrate concentrations at Eh values of approximately $+0.28$ volts. While these redox conditions are sufficient to cause denitrification in the aquifer, here denitrification could also occur in the clayey subsoils.

Mismanagement of both fertilizer and irrigation water in other highly vulnerable ground-water areas has caused or is the suspected cause of nonpoint nitrate contamination elsewhere in Nebraska. Studies in Holt, Cheyenne, and Garden Counties and in Hall County near Grand Island typify the occurrence of nonpoint contamination (Fig. 14).

Approximately 46,600 hectares of northern Holt County are underlain by nitrate-contaminated ground water (Exner and Spalding, 1979). This ground water also is characterized by elevated concentrations of chloride and sulfate which are added to the soils as sulfamag and potash to correct soil deficiencies. Rates of increase of 0.4 mg/L, 1.1 mg/L, and 0.6 mg/L were calcu-lated for chloride, nitrate-N, and sulfate, respectively, by plotting the ion concentration against the age of the irrigation well. These rates of increase were statistically significant at the 1% level. A comparison of the slope of nitrate to chloride in the ground water under ferti-lized and irrigated fields and the slope of nitrate to chloride in applied fertilizer suggested that 50% of the applied nitrogen fertilizer had leached to the ground water (Exner and Spalding, 1979).

The Sidney area of Cheyenne County and Oshkosh area of Garden County are relatively small areas of nonpoint source contamination; yet, they are representative of conditions in narrow alluvial valleys found throughout the Midwest and Great Plains regions. Stable nitrogen isotopes demonstrated that the source of most of the contamination near Sidney was leachates from manure applied to corn fields (Bryda, 1988), while near Oshkosh the source was leachates from commercial fertilizers (Exner, 1990).

Application of excessive sludge, and most probably excessive irrigation water, to well-drained bottomland soils near Grand Island resulted in a plume of nitrate contamination originating at the sludge injection site (Spalding, 1989). Nitrogen isotope measurements indicated the source was primarily animal waste and/or domestic sewage sludge.

Deep soil coring beneath excessively fertilized, irrigated research plots revealed that slugs with significant nitrate concentrations had moved downward at least 18 meters in 15 years (Spalding and Kitchen, 1988). Research has yet to be completed on the movement of nitrate beneath dry land agriculture.

In the glaciated areas of eastern Nebraska, elevated nitrate levels appear to be associated with farmhouse-barnyard complexes. Exner et al. (1985) concluded that almost half of the sampled wells in the glacial drift of southeastern Nebraska (Fig. 11) were poorly sited near potential point sources of nitrate contamination, namely, intermittently used and abandoned barnyards, and that the large number of poorly constructed wells increased the incidence of contamination. Some of the nitrate in the shallow alluvial valleys of the glacial drift region is fertilizer leachate, although the resulting concentrations seldom exceeded the MCL (Tanner and Steele, 1991).

About 60% of Kansas is cropland (U.S. Department of Commerce, 1989). The nation's wheat capital, Kansas also leads the country in the production of grain sorghum. In Madison and Brunett's (1985) data compilation, Kansas had the highest incidence of nitrate concentrations above 3 mg/L NO_3-N (Fig. 1). Twenty percent of the concentrations exceeded the MCL. In a recent, randomized, statewide survey, Steichen et al. (1988) found 28% of the 103 farmstead wells sampled exceeded the MCL. Concentrations were most likely to exceed the MCL in the northern tier of counties along the Kansas-Nebraska border and in central Kansas. Although the sample base was too small to determine sources, the low frequency of pesticide detections (only 4% of the samples contained atrazine) suggested that the elevated concentrations were far from exclusively nonpoint in origin.

In lysimeter studies beneath chemical flooding experiments in central Kansas counties, Sophocleous et al. (1990) found irrigation and fertilizer application rates and movement of nitrate were closely associated. Increases in the chloride to nitrate ratios in the shallow aquifer system were seen as evidence of denitrification.

water quality study, Division of Water and Natural Resour Management, Pierre

Spalding RF (1989) Ground water quality as influenced by sludge application at Grand Island, Nebraska, Contract Rept, Water Center, Institute of Agriculture and Natural Resources, U of Nebraska, Lincoln

Spalding RF (1990) Water quality in the Lower Platte River Basin with emphasis on agrichemicals, Contract Rept, Conservation and Survey Division, Institute of Agriculture and Natural Resources, U of Nebraska, Lincoln

Spalding RF, Exner ME, Lindau CW, Eaton DW (1982) Investigation of sources of groundwater nitrate contamination in the Burbank-Wallula area of Washington, U.S.A. J Hydr 58:307-324

Spalding RF, Kitchen LA (1988) Nitrate in the intermediate vadose zone beneath irrigated cropland. Ground Water Monitoring Review 8 (2):89-95

Steichen J, Koelliker J, Grosh D, Heiman A, Yearout R, and Robbins V (1988) Contamination of farmstead wells by pesticides, volatile organics, and inorganic chemicals in Kansas. Ground Water Monitoring Review 8(3):143-160

Stewart BA, Viets FG Jr, Hutchinson GL, Kemper WD, Clark FE, Fairbourn ML, Strauch F (1967) Nitrate and other water pollutants under fields and feedlots. Envr. Sci. Tech. 1:736-739

Stout PR, Burau RG (1967) The extent and significance of fertilizer buildup in soils as revealed by vertical distribution of nitrogenous matter between soils and underlying water reservoirs. In: Brady NC (ed) Agriculture and Quality of Our Environment, p 283

Tanner DQ, Steele GV (1991) Groundwater quality in the Nemaha natural resources district, southeastern Nebraska, USGS Water Res Invest Rept 90-4184

U.S. Dept. of Agriculture (1988) 1988 Agricultural Chartbook. USDA Handbook 673, Washington, D.C.

U.S. Dept. of Commerce (1989) 1987 Census of Agriculture Vol.1, Part 51, Washington, D.C., p 144

U.S. Environmental Protection Agency. (1990) National survey of pesticides in drinking water wells. Phase I report, EPA 570/9-90-015, Office of Water, Office of Pesticides and Toxic Substances, Washington, D.C.

Viets FG Jr, Hageman RH (1971) Factors affecting the accumulation of nitrate in soil, water and plants. U.S. Department of Agriculture - Ag Res Serv Agriculture Handbook No 413

Ward PC (1970) Existing levels of nitrates in waters - the California situation. In: Nitrate and Water Supply: Source and Control. Twelfth Sanit. Engin. Conf. Proc., Engin. Pub. Off., U Illinois, Urbana

Wells ER, Krothe NC (1989) Seasonal fluctuation of $\delta^{15}N$ of groundwater nitrate in a mantled karst aquifer due to macropore transport of fertilizer-derived nitrate. J Hydrology 112:191-201

THE NITRATE CONTENT OF DRINKING WATER IN PORTUGAL

S.M. Cardoso
Institute of Hygiene and Social Medicine
University of Coimbra
3049 Coimbra Codex, Portugal

Abstract

Nitrate levels in drinking water are rising, due in part to demographic growth and also because of agricultural development. Nitrate levels in most Portuguese waters (85.3%) are below 30 mg/L, and below 10 mg/L in over 50% of the waters. However, levels are very high in certain regions (Ribatejo and Alentego), and in some cases reach 120 mg/L.

1 Introduction

Nitrogen is indispensable for protein synthesis by autotrophic organisms. When nitrogen is supplied to plants their growth is more rapid and intense. This central fact has informed the general use of nitrogen-based fertilizers in agriculture. Nitrate, of course, is one of the principle sources of nitrogen.

If on the one side there is a remarkable gain in plant development, on the other side we risk ingesting great amounts of nitrate either in food or water. In fact, the nitrate rate in drinking water has increased significantly in the last twenty years, and the concentration of nitrate in surface waters is today six times higher than twenty years ago (World Health Service (WHO), 1984).

2 Sources and Consumption of Nitrogen

The principal sources of nitrogen in water are (1) the fixation of atmospheric nitrogen by algae, (2) rain water, (3) household wastewater, and (4) chemical and "natural" fertilizers. Industrial wastewater is also an important source of nitrate, especially from cellulose factories. Demographic and industrial growth and the development of agriculture are mainly responsible for nitrate pollution in the rivers and sheet waters of Portugal.

Consumption of nitrogenous nutrients in Portugal has increased dramatically; indeed, by

NATO ASI Series, Vol. G 30
Nitrate Contamination
Edited by I. Bogárdi and R. D. Kuzelka
© Springer-Verlag Berlin Heidelberg 1991

1985/1986 consumption had increased by more than 70% as compared to the base year of 1961/1962 (Fig. 1). Portugal consumes 120 to 140 tonnes of nitrogenous nutrients each year (Fig. 2), and despite the great capacity to metabolize and use these products, nitrate enrichment is occurring, especially in surface waters.

Consumption of fertilizers varies from region to region (Fig. 3). Alentejo consumes more nutrients (34%) than the Midland (26%) or the North (21%), and consumption is lowest in the Algarve. Hence we may deduce that the risk of contamination will be higher in Alentejo than in other regions.

Concerning the nitrate concentration of drinking water generally, Portugal occupies an inter-mediate position. A study of 200 water samples drawn from throughout the country found that the nitrate concentration was below 10 mg/L in 52.6% of the samples (Fig. 4). However, nitrate levels in some regions, and especially in Ribatejo and Alentejo, exceed recommended WHO limits, and in some cases reach 120 mg/L. It is exactly in this region that more nitrogenous nutrients are consumed (Cardoso, 1983). However, drinking water is not the only source of nitrate; many products (e.g., vegetables) are also important sources of nitrate.

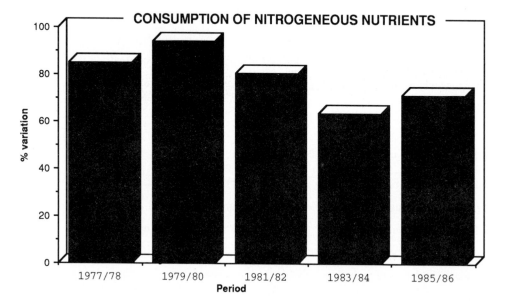

Fig. 1. Consumption (percent) of nitrogenous nutrients in Portugal (base = 1961/1962)

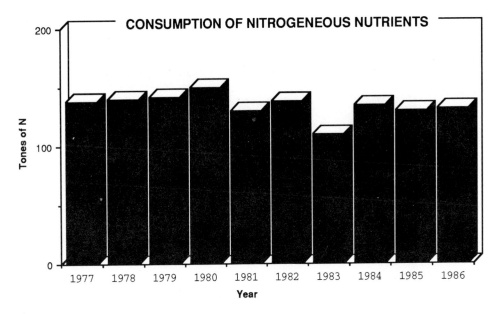

Fig. 2. Consumption of nitrogenous nutrients (tonnes of N/year)

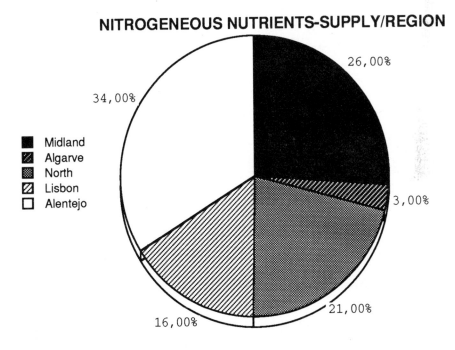

Fig. 3. Distribution (percent) of nitrogenous nutrients by region of Portugal (1985/1986)

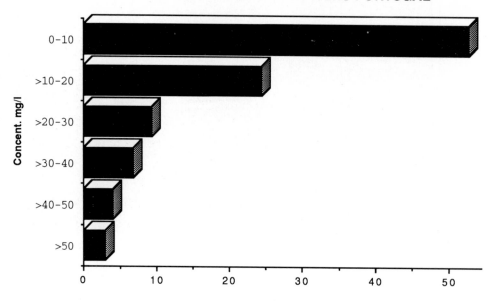

Fig. 4. Distribution of drinking water in Portugal by nitrate rate

Table 1. Food nitrites before and after Ultra Violet Radiation (U.V.R.) (Cardoso, 1986)

Food	Nitrates and Food (Effect of U.V.R.) (mcg/g)		
	Basal	1 Hr.	%
"Corned beef"	17.5	31.5	+79.7
Ham	20.8	23.0	+10.7
Smoked Sausage	15.5	15.5	
Gammon	3.7	15.2	+310.8
Large sausage	5.5	5.5	
Pressed meat	2.1	5.2	+147.6
Sausages	1.4	4.6	+228.9
Narrow sausage	1.3	2.1	61.5
Blood sausage	0.0	0.0	
Flour sausage	0.0	0.0	

Cardoso et al. (1986) studied nitrite levels in several food products. Some food products (Table 1) did not show either nitrites or N-nitroso compounds, and therefore cannot be considered dangerous (blood sausage and flour sausage); others had low rates. Among these, the nitrate concentration of narrow smoked sausage, which presented 1.3 mcg/g of nitrites in basal conditions, rose to 2.1 mcg/g after being submitted to U.V.R. (an increase of 61.5%). The rate for sausages rose by 228.6%, probably due to the presence of N-nitroso compounds. Nitrate concentrations in "corned beef" (imported) and gammon also rose considerably.

Nitrite and nitrosamines can also be preformed (Cardoso, 1986). The formation of nitroso-amines was observed *in vitro* starting from nitrites and from secondary amines after an incubation with the gastric contents of different animal species. Long-term administration of amines and nitrites in drinking water caused a neoplasma in a mouse (Martin, 1979). Other researchers consider that nitrites in the gastric element are primarily salivary in origin because of the reduction action of bacterial flora. Saliva is an important source of nitrites; in fact, the salivary glands behave in relation to nitrate just as the thyroid gland in relation to iodine (Cardoso, 1985).

In the gastric-salivary cycle, nitrates concentrated in the salivary glands reduce to nitrites; most are subsequently reabsorbed. Most of the nitrates which are of hydro-nutritional origin are eliminated in the urine.

Production of salivary nitrites is affected by the nitrate concentration in the water supply. Concentration of salivary nitrates in normal subjects increased on average up to 3.2 times the original level 1 hr after ingestion of 12.5 mg of potassium nitrate in water (Cardoso, 1985).

3 Conclusion

There is no doubt that the cancerogenic action of nitrosamines is responsible for several neoplasias (liver, esophagus, stomach, lungs, kidneys, pancreas, bladder, central nervous system, etc.). Over one hundred of these compounds have been identified, and most of them are undoubtedly carcinogenic.

Although food contains higher levels of nitrate than water, water-based nitrate contamination must be considered in reduction of cancer risk. The increased use of fertilizers is part of the

for the last two years. The second study deals with ground waters in the Attiki area, near Athens, where high nitrate levels have been detected. Various sources of nitrate contamination are considered, as well as the range of technical alternatives which are available to prevent nitrate pollution of ground water. These techniques can be more or less efficient depending on the degree of knowledge of the existing environmental situation, as well as of the relation between the pollutant loads from external sources and the nitrate concentration in the water. Mathematical modeling is a very useful tool for assessing the state of pollution in the water environment and for predicting the efficiency of remediation alternatives. Numerical techniques which were developed in the Hydraulics Laboratory at the Aristotle University of Thessaloniki (AUT) for modeling the transport and fate of nitrate in the water environment are briefly discussed.

2 Present Situation

The morphology of the hydrologic basins in Greece, which are generally small and steep, does not favor the penetration of nitrate into ground-water aquifers. Nitrate levels in Greek waters seem to be generally rather low in relation to the relatively nonintensive use of fertilizers in agriculture; however, some of the available data show that in many cases there is a systematic upward trend in these levels.

Although the available data do not provide a complete picture of nitrate contamination of Greek waters, an attempt to summarize the present situation is provided in Fig. 1. The probability is very high that nitrate levels will continue to increase in the Pinios (Thessalia), Axios (plain of Thessaloniki), and Nestos (plain of Serres) Rivers. This is also the case in the Bay of Amvrakikos (western Greece, Fig. 1) and Lake Visthonis (eastern Greece, Fig. 1); and nitrate levels are also high in the ground-water aquifer in the Attiki area (Fig. 1).

In fact, fertilizer consumption in Greece has steadily increased over the last years, following similar international trends. Fig. 2 shows the consumption of nitrogen (N) and commercial (NPK) fertilizers in Greece during the period 1970-1983 (OCDE, 1985).

2.1 Axios River

Wastewater from the greater Thessaloniki metropolitan area is released to the Axios River

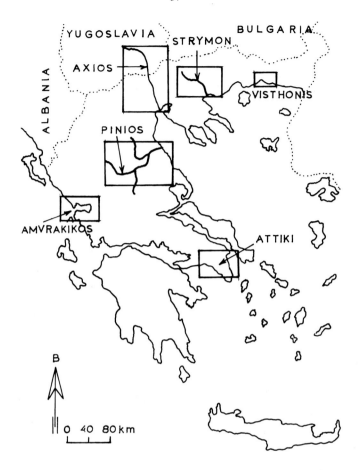

Fig. 1. Areas of high risk for water contamination by nitrate in Greece

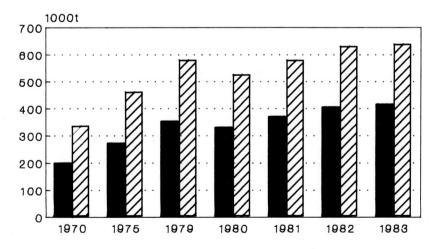

Fig. 2. Nitrogen (N) and commercial (NPK) fertilizer (x 10^3 t) consumption in Greece (1970-1983)

after treatment. Concern for the possible effects of this decision on water quality led to the establishment in 1988 of an extensive monitoring program by the Hydraulics Laboratory at AUT. The program is intended to supplement data collected by the Ministry of Agriculture.

As shown in Fig. 3, the Axios River flows for approximately 75 km between the Greek-Yugoslavian border and the sea. This corresponds only to 10% of the total basin area of 23,750 km^2, with the remainder located in Yugoslavia. The mean flow rate of the river is about 170 m^3/s, but falls to a minimum of 37 m^3/s during the summer. Samples have been collected on a monthly basis at monitoring station III since April 1988 (Fig. 3). The following water quality parameters have been analyzed: temperature, pH, dissolved oxygen, BOD, COD, suspended solids, salinity, conductivity, nitrites, nitrate, ammonia, total organic nitrogen (NK), phosphates, silicates and heavy metals.

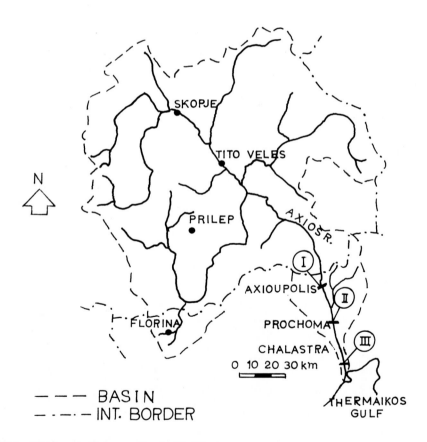

Fig. 3 Monitoring stations on the Axios River

Time series data for nitrogen-related parameters are shown in Fig. 4. The values (in ppm) are rather low and, in particular, are far below prescribed drinking water standards. Comparison of the mean nitrate-nitrogen concentrations with corresponding values for the period from 1981 to 1982 (Table 1) shows that in a period of only seven years nitrate-nitrogen concentrations increased by almost 50%.

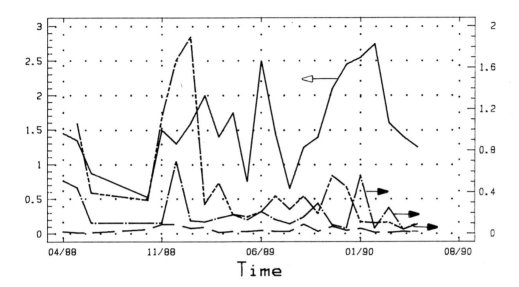

Fig. 4. Time series of nitrogen-related parameters at Station III (Axios River)
(— NO₃-N   -- NO₂-N   — · — NH₄-N   — ·· — NK)

Table 1. Values of NO₃-N concentrations (in ppm) in the Axios River

Period	Minimum	Mean	Maximum
1981-82	0.50	1.05	1.76
1988-90	0.52	1.56	2.75

This trend was confirmed for the period 1988 to 1990 by the linear regression shown in Fig. 5. Exploring possible sources for nitrate pollution in the river Axios, the data were analyzed for seasonal variation. In a computation of the autocorrelation coefficient of the nitrate time series, no significant correlation was found for time periods longer than three months. When the seasonal subseries of NO_3-N concentrations are plotted over a one-year

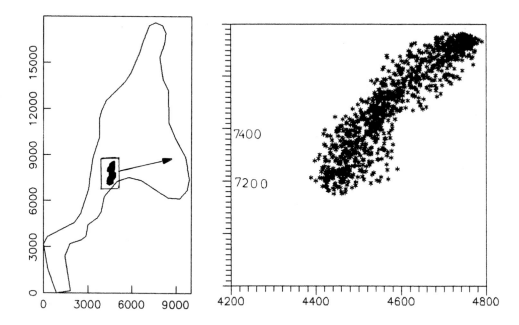

Fig. 9. Use of random-walk simulation for the study of ground-water pollution of the Gallikos aquifer (dimensions in meters)

Mathematical modeling is a useful tool to improve application of remedial measures. Some of the numerical-simulation techniques which can be used to predict surface and ground-water contamination by nitrate are summarized in this paper. The techniques have been developed in the Hydraulics Laboratory at AUT.

References

Giannakopoulou T (1990) Reliability of some eutrophication indices in an estuary (in Greek). In: Proc. 4th Congress of the Greek Hydrotechnical Union, Heraklion, p 675

Jobson HE (1987) Lagrangian model of Nitrogen kinetics in the Chattahoockee River. J Environ Engr Div ASCE 113:223-242.

Katsiri A (1983) Groundwater pollution of Attiki area by nitrates (in Greek). In: Proc. 1st Congress of the Greek Hydrotechnical Union, Thessaloniki, II.85

Latinopoulos P, Ganoulis J, Tolikas D (1982) Efficiency of integral equation method for modelling regional groundwater flows. In: Proc IAHR Int'l Conf on Groundwater Resources Management, Capri, Italy.

OCDE (1985) Data on the environment. Compendium OCDE, Paris

Starr JL, Broadbent, FE, Nielson DR (1974) Nitrogen transformations during continuous leaching. Soil Sci Soc Am 38:283-289.

Thomann RV, O'Connor DJ, DiToro DM (1971) The effect of nitrification on dissolved oxygen of streams and estuaries, Manhattan College, Bronx, Environmental Engineering and Science Program Technical Report

II. EXPOSURE ASSESSMENT

RISK ANALYSIS FOR WATER SUPPLY FROM A RIVER POLLUTED BY NITRATE RUNOFF

E.J. Plate and L. Duckstein[1]
Institute for Hydrology and Water Resources Planning
Karlsruhe University, 7500 Karlsruhe, Germany

Abstract

A stochastic model is developed for use in determining the probability of nitrate pollution in the drinking water drawn from a river loaded by drainage from fertilized fields. Both the runoff of nitrate and the river discharge are random variables whose combination yields the desired probability. The problem is formulated in terms of reliability theory, and the model is formally extended to cover both input variable uncertainty and risk assessment. The risk can be used as a decision variable for the selection of remedial measures.

Although the model is only conceptual, it is the first step in an extensive effort to produce a decision model for pollutant control from agricultural fields. In order to ascertain which model concepts will be further developed, a project has been initiated at the University of Karlsruhe for an in-depth study of the event chain starting with the application of nitrate-based fertilizers and ending with river nitrate concentrations. The study proceeds experimentally through intensive field studies, and theoretically through construction of models which will be calibrated with data from the field study. A brief description of the study is given.

1 Introduction

The purpose of this paper is to present a conceptual model of nitrate pollution from agricultural fields for use in estimating the joint risk from field runoff loading and from concentrations present in a river flowing through the area. The model is linked to a major field study being undertaken near Karlsruhe, Germany.

Nitrate runoff from agricultural fields has been identified in many areas as the major source of overfertilization of receiving waters, which leads to algae bloom and nitrate accumulation

---

[1] Department of Systems and Industrial Engineering, University of Arizona, Tucson, AZ, U.S.A.

NATO ASI Series, Vol. G 30
Nitrate Contamination
Edited by I. Bogárdi and R. D. Kuzelka
© Springer-Verlag Berlin Heidelberg 1991

in drinking water supplies. Spurred by the need for an adequate return from agricultural areas, fertilizer use has increased dramatically. In (former) West Germany, for example, application of mineral nitrate fertilizers increased from 25.6 kg/ha to 112.5 kg/ha between 1950 and 1980, and organic fertilizer application from 31.9 kg/ha in 1950 to 73.3 kg/ha in 1980 (Kretzschmar et al., 1985). During the same period, wheat production rose from 2.6 to 5.0 metric tons per hectare (t/ha). As a result, nitrate runoff from ground and surface water has increased.

Policies to reduce fertilizer use have been proposed, including restrictions on the maximum amount of fertilizer which may be used in water-protection areas (areas where ground water is used for drinking water), or voluntary reductions in the amount of fertilizer applied to fields. A political debate has begun over the question of who should pay for the cost of such preventive measures. For example, the State of Baden-Württemberg has shifted from the "polluter principle" to collecting user's fees from waterworks customers to use to compensate farmers for reduced production. The constitutionality of this legal novum has not yet been reviewed by the German Supreme Court.

Mathematical models are needed to assess nitrate pollution and the effects of localized nitrate-reduction measures, as well as to estimate the cost of the pollution-reduction actions. Such models must be able to quantify adequately the process of nitrate transport and nitrate depletion and use. In particular, a purpose-oriented decision model is needed to determine if the water drawn from a river at a point downstream of the fertilized area meets quality standards expressed in terms of a permitted nitrate concentration c_p. That is, is the permitted nitrate concentration c_p exceeded sufficiently frequently to cause nitrate problems? This paper presents the framework of a stochastic model developed for this purpose. The model is based on reliability concepts presented in Duckstein and Plate (1987), Plate and Duckstein (1988), and Plate (1986).

2 A Stochastic Nonpoint Pollution Model for Rural Areas

A stochastic nonpoint pollution model must consist of three parts. The first is the input model. Inputs consist of rainfall fields, which provide the runoff for the pollutant transport, and fertilizer inputs for each field, identified as M_i for the i-th field. The input model also includes process models which convert rainfall into runoff and fertilizer mass into fertilizer concentration c_i in the runoff from field i. The input model provides the loading into the river. The second part is a transport model specifying transport and mixing in the river. This second part is also a process model, because it incorporates transport by convection and by diffusion between the point of discharge from each field i into the river and the point of water withdrawal from the river. This model yields a time series of the concentration $c(t)$ and of discharge $Q(t)$ in the river at the withdrawal point. The third part, called the decision model, expresses the objectives which should be satisfied by the stochastic model.

In its most elementary form, the stochastic model consists of mass-balance relationships with statistical parameters, and decisions at any level will thus have to be made on a statistical basis. The problem associated with decision making on the basis of nitrate concentrations in a river may then be taken as a design problem based on failure probability, to which standard techniques of reliability analysis (Duckstein and Plate, 1987) or of stochastic design (Ang and Tang, 1984; Plate and Duckstein, 1988) can be applied. Let us consider the three models as they are integrated into a stochastic simulation procedure.

2.1 The Input Model

We begin by looking at a situation involving contaminant runoff from a rural area into a river coming from an unpolluted area; the concentration in the river is zero before it enters the area (Fig. 1). We assume that the rural area consists of many different fields, each of which is

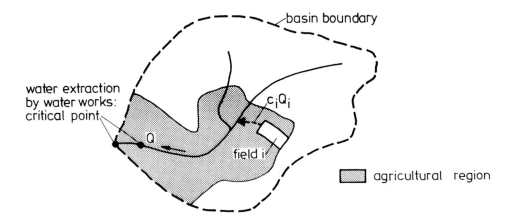

Fig. 1. Definition of the problem considered

situation in which the standards are not met shall be called a "failure" (i.e., if the river or creek is in the state $c > c_p$). A reliability-based decision model starts from the assumption that it is not necessary to completely avoid the "failure" condition, but that intolerable damage (i.e., health hazards or damage to the ecosystem) will occur only if the failure probability $P_F = P\{c > c_p\}$ is higher than some critical value.

The failure condition $c > c_p$ is a stochastic event. In terms of the stochastically independent variables Q and M, the event can be rewritten in order to separate the two random variables river discharge Q and mass $M = c\,Q$ of nitrate in the river. This is done by multiplying the failure condition with Q. Thus we obtain the inequality in terms comparing the injected mass M with the theoretically permissible mass $c_p\,Q = M_p$:

$$M > c_p\,Q \qquad\qquad [4]$$

The left side of Equation [4] is a random variable because of concentration runoff from fields; the right side is a random variable because of variability of discharge Q. The model expressed by Equation [4] is equivalent to the classical case of calculating the probability of failure for random resistances and loads, $r = c_p\,Q$ and $s = M$, as illustrated in Fig. 2. Fig. 2 shows the

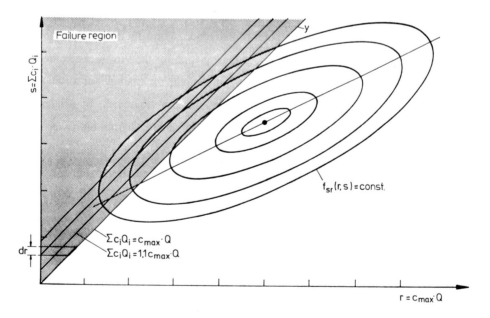

Fig 2. Definition of the probability that concentration c in the river exceeds a critical value c_p

two-dimensional probability density distribution $f_{rs}(r,s)$ for resistance and load, in which it is assumed that statistical dependence exists between both variables. Marginal distributions $f_r(r)$ and $f_s(s)$ are also plotted on the axes. Except for the scaling factor c_p, $f_r(r)$ is identical to the probability density of Q, whereas $f_s(s)$ is the density of nitrate mass loading into the river.

In this diagram, condition $r=s$, which corresponds to the failure surface, is a straight line bisecting the $r - s$ quadrant. The probability of failure is that of finding a combination of r and s in the shaded region of the $r - s$ quadrant; it is calculated from the probability distribution of pollutant excess z defined as:

$$z = s-r \qquad [5]$$

with mean value $\mu_z = \mu_s - \mu_r$.

The equation for f_z (f) can be obtained by integration according to Fig. 2, using the formula (Papoulis, 1965):

$$f_z(z) = \int_{-\infty}^{\infty} f_{rs}(s^*,r) \cdot \left| \frac{ds^*}{dz} \right| \cdot ds \qquad [6]$$

where s^* is the transformation of $s(z)$ for condition $z=$constant. In our case s^* is simply:

$$s^* = z + r \qquad [7]$$

Consequently we obtain:

$$f_z(z) = \int_{-\infty}^{\infty} f_{rs}([z+r]|r) \cdot f_r(r) \cdot dr \qquad [8]$$

where $f_{rs}(R|S)$ is the conditional probability density for r given s. In general, a straightforward solution of Equation [8] cannot be obtained. However, a solution is available if we can make the assumption that Q also follows a normal distribution, and that variables r and s have a correlation coefficient ρ. Then the joint probability distribution of c/c_p and Q is described by:

NITROGEN MODELING ON A REGIONAL SCALE

P.E. Rijtema and J.G. Kroes
The Winand Staring Centre for Integrated Land, Soil and Water Research
P.O. Box 125
6700 AC Wageningen, the Netherlands

Abstract

Processes which are involved in nitrogen modeling on a regional scale are discussed, and the performance of some models in relation to measured field data is analyzed. Data on nitrate-N emission for different fertilizer strategies are simulated on a national scale, and cost comparisons for different control strategies are provided in relation to nitrate-N emission reduction. Uncertainties in simulation results as well as in measured field data are discussed. It is concluded that the simulation model is realistic and meaningful, and provides a good basis for policy analysis.

1 Introduction

Formation and implementation of a rational and effective policy to reduce the nitrogen load on ground and surface waters requires a thorough understanding of the behavior of nitrogen and its compounds in the soil. This understanding can be increased by using simulation models to predict leaching of nitrates through the soil to the ground water for a variety of soil types, climates and hydrological conditions, and using different scenarios for agricultural development. Many issues need to be addressed:

1. Translation of point and field studies to the regional and even national scale
2. An exact consideration of the short and long-term effects of pollution and its control in relation to the reaction time of the system
3. A more comprehensive consideration of uncertainties in the control objectives and alternatives; and the structure, parameters and input data of the applied models
4. Consideration of the uncertainties in model calibration and validation

The processes which are involved in nitrogen modeling and the performance of several models are discussed. Results for different fertilizer strategies as simulated with the ANIMO model on a national scale are presented, followed by a discussion of the uncertainties in the simulation results.

NATO ASI Series, Vol. G 30
Nitrate Contamination
Edited by I. Bogárdi and R.D. Kuzelka
© Springer-Verlag Berlin Heidelberg 1991

2 Nitrogen Models

There is a need to quantify the sources of nitrogen compounds from rural regions under various climate and soil conditions, water management systems, cropping patterns, and fertilizer type and application technology. Technological changes in agricultural production should be evaluated in terms of their effects on fertilizer use and their environmental impacts.

Nitrogen in the soil primarily originates from inorganic and organic fertilizers, organic matter, precipitation and irrigation water. It is mainly removed by crops, drainage water, denitrification and volatilization of ammonia. A suitable model for regional use should be based on a clear and quantitative description of the following main processes:

1. Mineralization and immobilization of nitrogen related to processes in the carbon cycle
2. Denitrification related to (partial) anaerobiosis and decomposing organic materials, which implicates the modeling of the oxygen and temperature distribution in the soil and the transport of various compounds

These processes take place in two zones, the upper unsaturated zone and the lower saturated aquifer. The conditions in the upper unsaturated zone are strongly determined by infiltration and also by the soil moisture distribution in the unsaturated zone. Soil moisture conditions determine the aeration in the unsaturated zone, and this is directly related to the mineralization of organic nitrogen as well as to the denitrification rate. Although processes and transport of solutes in the unsaturated zone can be described one-dimensionally, this is not the case in the saturated subsoil where transport to different drainage systems also should be considered.

Many models have been developed simulating nitrogen behavior in the soil, ranging from scientific research tools to simple management models. An extensive comparison of the concepts and performance of a number of models was carried out by Frissel and Van Veen (1981) using different data sets for each model.

Hardly any comparisons of nitrogen leaching models are available that use the same data sets and employ quantitative criteria to evaluate model performance. Reiniger et al. (1990) presented selected results of a preliminary study of leaching models. Based on a first screening of their structure, concept process description and development stage, these authors selected five of ten available models for a comparative evaluation. These models are deterministic and one-dimensional, and they can be regarded as functional models carrying out dynamic simulations. They can be differentiated conceptually on the basis of their description of water and solute flow and nitrogen transformation. The models ANIMO (Rijtema et al., 1991), LEACHM (Wagenet and Hutson, 1989), RENLEM (Kragt et al., 1989) and SWATNIT (Vereecken et al., 1989) use the Richard equation to calculate the components of the soil water balance, and they can handle the effects of varying ground-water tables. EPIC (Williams et al., 1983) uses a capacity-type model.

ANIMO, LEACHM and SWATNIT describe the complete cycle of nitrogen and carbon transformation processes. EPIC describes part of the nitrogen transformation process and does not consider the carbon cycle. RENLEM describes part of the nitrogen transformation process and uses a simplified approach for the carbon availability, assuming steady-state conditions in the soil system.

ANIMO uses a two-dimensional transport module in the saturated zone to calculate solute transport to different drainage systems. The data reported by Reiniger et al. (1990) (Fig. 1) indicate considerable differences between calculated and measured nitrate concentrations.

3 Model Application on a Regional Scale

3.1 Background

In the context of the preparation of the third Policy Analysis of Water Management in the Netherlands (PAWN), the effects of different scenarios of land use development and fertilizer application on the nutrient load on ground and surface waters were analyzed. The following issues were addressed:

1. The present situation concerning the nutrient contribution from agricultural lands to surface runoff, interflow and deep drainage flux

2. The situation predicted for the year 2000 considering a number of scenarios for agricultural development involving restrictions on the use of animal slurries, application technologies and two different European Community (EC) price policies for agricultural products

3. The additional costs of the different scenarios

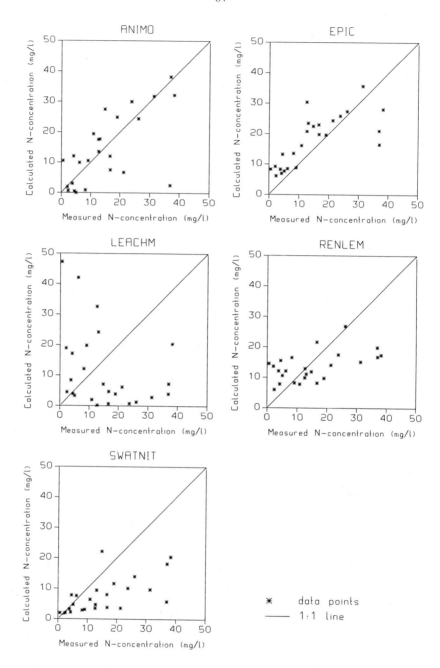

Fig. 1. Measured and calculated nitrate-N concentration at one meter below soil surface, averaged over the winter season for five simulation models (after Reiniger et al., 1990)

The model study was subject to the following boundary conditions:

1. The study had to use the results of Grashoff et al. (1989) as input for the levels of N and P fertilizer application.

2. The output of the hydrological model DEMGEN had to be used for the water fluxes to different drain systems as input in the model ANIMO, using the defined standard soils of the hydrological analysis (Grashoff et al., 1989).

The results were to a great extent determined by these boundary conditions. The model study was performed with an extended version of the ANIMO model, which also included a module for the P cycle in soils (Rijtema et al., 1991).

3.2 Schematization

The starting point of the PAWN 1 study (Abrahamse et al., 1982), was the areal schematization of the Netherlands into a number of districts based on major water-supply units and drainage catchments. The districts were further subdivided into geographically undetermined subdistricts having only one standard soil profile and one predescribed set of hydrological parameters. The subdistricts, in turn, were subdivided into plots on the basis of land use. In this way the whole country was subdivided into 77 districts, 143 subdistricts and 503 plots, using 26 different standard soils. Four major types of land use were considered: *grassland*, *forage maize*, *arable land with crop rotation*, and *other land use* (consisting of nature reserves, forests, small open waters and gardens). Urban areas and main open waters were not considered in the study. The different land use forms in the Netherlands are presented in Table 1. Input data used in the study are described by Kroes et al. (1990).

Table 1. Acreage of the different types of land use in the Netherlands. Urban areas and main open waters are excluded

Land Use	Acreage (ha)	%
Grassland	1,161,687	40
Forage maize	170,316	6
Arable land	596,068	21
Other land use	943,695	33
Total	2,871,766	100

PRINCIPLES OF MONITORING AND ANALYSIS

H.P. Nachtnebel, L. Duckstein,[1] and I. Bogardi[2]
Institute of Water Resources Management
Universität für Bodenkultur
Gregor Mendel-Str. 33
A-1180 Wien, Austria

Abstract

The aim of this paper is to evaluate the efficiency of ground-water monitoring networks with respect to several objectives. The efficiency is expressed by the standard deviation of the distribution of the predicted nitrate exposure. In the case of basin-wide nitrate monitoring, the spatial pattern of the nitrate concentration and the estimation variance are obtained by geostatistical analysis. In the second case, which refers to the impact assessment of distinct sources of pollution, a simulation technique is applied to estimate the estimation variances of nitrate. The simulation technique is based on a ground-water solute transport model which requires various hydrogeologic parameters. The information obtained serves as an input to the health risk assessment and relates the monitoring efficiency, the uncertainty in the exposure and the costs to health risk.

1 Introduction

Monitoring activities are an important step in the planning, management and protection of ground-water systems. Depending on the objectives of the monitoring program, different networks, sampling strategies and measurement techniques are applied. The objectives are either to continue surveillance and identification of trends and seasonal variations, or to monitor the system's reaction to different management alternatives. Monitoring methodologies were formulated and described by Tinlin and Everett (1978), LAWA (1982), and EPA (1985).

Ground-water quality is strongly dependent on hydrogeologic parameters, soil type, land use and water management (Steenvoorden, 1976; Rijtema, 1980; Canter and Knox, 1985; Schiffler, 1990; Jensen and Refsgaard, 1988).

---

[1] Systems and Industrial Engineering, University of Arizona, Tucson, AZ 85721, U.S.A.
[2] Civil Engineering, University of Nebraska-Lincoln, Lincoln, NE 68588-0531, U.S.A.

NATO ASI Series, Vol. G 30
Nitrate Contamination
Edited by I. Bogárdi and R. D. Kuzelka
© Springer-Verlag Berlin Heidelberg 1991

In this paper the emphasis is on the estimation of average nitrate concentrations and of the respective estimation variance, since these two figures provide important information for health risk assessment. Aspects of two types of monitoring networks, *ambient trend monitoring*, and *source assessment*, are discussed. The analysis of the sample data yields the spatial variations of pollution and the measurement (estimation) variance in consideration of the network geometry. Two examples, one illustrating a risk-based approach and the other a multiobjective framework, will be used to demonstrate the value of information obtained from the monitoring network.

2 Risk-based Approach for Network Evaluation

A common formulation of public health risk corresponding to an event M, say mortality due to cancer, is given by

$$R(M) = d \cdot p \cdot n \cdot T \qquad [1]$$

where n = the number of persons exposed;

d = the average individual dose, mg/kg.day;

p = the potency, which describes the slope of the dose-response relationship at low dose levels, and is expressed by death per person, mg/kg.day; and

T = the period of exposure in days (Anderson, 1983).

In reality, the true levels of d and p are unknown and can be only seen as realizations of random variables described by their respective (or joint) probability distributions. The expected value of risk $EV(R(M))$ is

$$EV(R(M)) = n \cdot T \int_{d_0}^{\infty} d(x) \cdot p\,(f(x))\,dx \qquad [2]$$

where d_0 is the level of "safe dose" or natural background concentration.

A monitoring system should minimize costs and simultaneously minimize the uncertainty in the evaluation of risk $R(M)$. These two conflicting objectives can be handled in a multiobjective framework (Kelly et al., 1987) or integrated into a single monetary objective function (Finkel and Evans, 1987), given as:

$$TSC = C_J + V(1 - e_j) \cdot R(M) \qquad [3]$$

where TSC = total social costs, TSC;

C_j = cost to implement and operate a control alternative, A_j;

V = social costs related to the risk, R;

R = the uncontrolled risk level, and

e_j = the risk reduction efficiency due to control alternative, A_j.

For instance, $V(R)$ may express the value of human life. The objective is to minimize total social costs. If the risk level is unknown, an appropriate decision will be to minimize the expected TSC corresponding to a certain alternative A^*. At a particular risk level R', another alternative A' may have a lower TSC. The respective opportunity loss OL is given by

$$OL(A^* | R') = TSC(A^* | R') \, f(R') \, dR' \qquad [4]$$

By integrating opportunity losses over the full range of risk levels the expected opportunity loss $EOL(A^*)$ is obtained:

$$EOL(A^*) = \int OL(A^* | R') \, f(R') \, dR' \qquad [5]$$

If perfect information about risk were available, the EOL would be zero. In other words, the value of EOL represents an upper bound on the value of information. Whenever the cost of acquiring new (additional) information exceeds the EOL, the collection of additional information would be a bad investment, even if perfect information would be gained.

In real-world problems, collection of information reduces uncertainty. It can be shown that the expected value of sample information, or imperfect information EVI, equals the difference in EOL prior to and after the additional sampling (Davis and Dvoranchik, 1971; Davis et al., 1972):

$$EVI = EOL_p - EOL_a \qquad [6]$$

where EOL_a = $\int\int OL(A^* | R) \cdot f''(R | S) \, f(S) \, dSdR,$

$f''(R | S)$ = $f(S | R) \cdot f'(R)/f(S),$

$f''(R | S)$ = posterior risk density given the sample information, S,

$f(S)$ = unconditional probability of observing such a sample result,

$f(S|R)$ = conditional probability of observing S given R, and

$f'(R)$ = prior risk density.

The value of decreasing the uncertainty in the dose increases with the level of uncertainty. The parameter Sp_i represents the standard deviation of the potency. The value of sample information increases as the other source of uncertainty, the potency p, decreases as can be seen from Fig. 1.

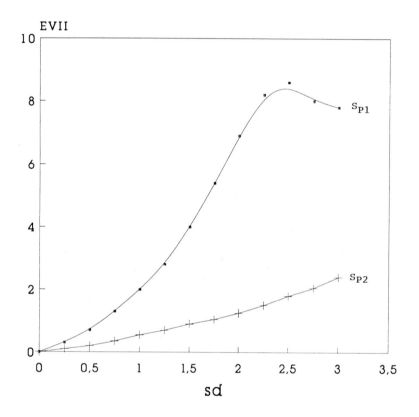

Fig. 1. Value of sample information *EVS* versus monitoring efficiency, S_d, and S_p (standard deviation of the dose and potency factor response, $S_{p1} < S_{p2}$)

3 Multicriterion Approach to Network Evaluation

In this approach the estimation variance in the dose (O_1) and the costs (O_2) are considered as separate objectives. Each network yields different information at different costs; the tradeoff between seven alternatives is sketched in Fig. 2. A monitoring alternative A_j can be selected from the candidate set of alternatives if weights are assessed to trade off the objectives. Kelly et al. (1987) compare combinations of two networks, one designed for geoelectric measurements and the second consisting of observation ground-water wells.

It can be observed in Fig. 2 that monitoring alternative 6 dominates alternative 7 because it results in a better outcome with respect to the cost objective and has the same standard deviation. If a compromise between network efficiency and costs has to be achieved, alternative 3 may be recommended. This alternative stands for a combination of 87 wells and ten geoelectric measurements.

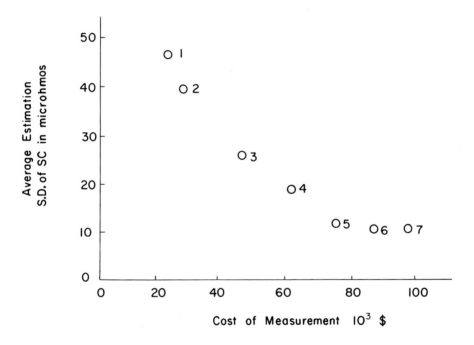

Fig. 2. Estimation uncertainty versus costs

4 Estimating the Uncertainty of the Exposure

In both the risk-based approach and the bio-objective case, the estimation variance of the exposure is essential for evaluating the efficiency of a monitoring network. In this section two different examples illustrate how the estimation variance can be obtained.

Todd (1976), and Canter et al. (1988) distinguish between two types of monitoring networks

From an engineering viewpoint, both monitoring and data analysis should address the spatial distribution of the nitrate pollution described by its expected value and estimation variance. Based on this information a rational evaluation of monitoring networks can be performed, and the effect of measuring specific parameters on the overall uncertainty in the exposure can be analyzed. Finally, the proposed approach assists in the design or extension of a network.

Acknowledgements

This research was partly funded by joint NSF-FFF research grants INT 8620200 and P 6502 B, respectively.

References

Anderson EL (1983) Quantitative approaches to use in assessing cancer risk. Risk Analysis 3:277-295

Bardossy A, Bogardi I, Kelly WE (1989) Geostatistics utilizing imprecise information. Fuzzy Sets and Systems 31:311-328

Brooker PI (1985) Simulation of spatially correlated data in two dimensions. Mathematics and Computers in Simulation 27:155-157

Canter LW, Knox RC (1985) Groundwater pollution control. Lewis Publ. Inc. Michigan, USA

Canter LW, Knox RC, Fairchild DM (1988) Groundwater quality protection. Lewis Publ. Inc., Michigan, USA

Cook DK (1981) Selection of monitoring well location in east and north Woburn, Massachusetts. In Proc. of the National Conference an Management of Uncontrolled Hazardous Waste Sites. Hazardous Materials Control Research Institute, Silver Springs, Maryland, p 63-69

Davis DR, Kisiel CC, Duckstein L (1972), Bayesian decision theory applied to design in hydrology. Water Resour Res 8(1):33-41

Davis DR, Dvoranchik WM (1971) Evaluation of the worth of additional data. Water Resour Bull 7(4):700-707

EPA (1985) Groundwater monitoring strategies. U.S. Environmental Protection Agency, Washington, D.C.

Finkel AM, Evans JS (1987) Evaluating the benefits of uncertainty reduction in environmental health risk management. JAPCA 37:1164-1171

Huijbregts CJ, Matheron G (1971) Universal kriging. Decision Making in Mineral Ind, CIM. Spec. Vol. 12:159-169

Isaaks EH (1985) Probability kriging. Report from Applied Earth Science Dept., Stanford Univ. CA

Jensen KH and Refsgaard JC (1988) Spatial variability of physical parameters and processes in field soils. Proc. Nordic Hydrological Conference, Part II, pp. 121-138, pp. 139-154, pp. 156-168. Helsinki, Finland

Journel AG, Huijbregts CJ (1978) Mining Geostatistics. AP, London, UK

Kelly WE, Bogardi I, Bardossy A (1987) Evaluation and design of combined networks for groundwater pollution monitoring. In: Van Duijveyboden W, Van Waegeningh HG (eds) Vulnerability of soil and groundwater to pollutants. Internat. Conference Nordwijk van See. Proc. and Information No. 38, p. 215-224, TNO Committee on Hydrological Research, The Netherlands

Konikow LF, Bredehoeft JD (1978) Computer model for two-dimensional solute transport and dispersion in groundwater. US Geological Survey, Reston, VA

LAWA (Länderarbeitsgemeinschaft Wasser) (1982) Grundwasser - Richtlinien für Beobachtung und Auswertung. Vlg. Paul Parey, Hamburg, BRD

Nachtnebel HP, Bardossy A (1990) Network geometry and spatial uncertainty in groundwater solute transport modeling. In Proc. of MODEL CARE, Den Haag, Netherlands

Neuman SP, Jacobson EA (1984) Analysis of nonintrinsic spatial variability by residual kriging with applications to regional groundwater levels. Math Geology 16(5):499-521

Omre H (1987) Bayesian kriging; Merging observations and qualified guesses in kriging. In Math Geology 19:25-39

Pfannkuch HO, Labno BA (1976) Design and optimization of groundwater monitoring networks for pollution studies. Groundwater 14(6):455-462

Rijtema PE (1980) Nitrogen emission from grassland farms - a model approach. Techn Bull 119, Institute for Land and Water Management Research, Wageningen, The Netherlands

Schiffler GR (1990) Design of soil moisture network in a small rural catchment. Proc. Hydrological Research Basins and the Environment. Wageningen, The Netherlands

Steenvoorden JH (1976) Phosphate and biocides in groundwater as influenced by soil factors and agriculture. Techn. Bull. 97, Institute for Land and Water Management Res., Wageningen, The Netherlands

Tinlin RM, Everett LG (1978) Establishment of groundwater quality monitoring programs. Proc. of AWRA Symposium, Minneapolis, MN

Todd DK (1976) Monitoring groundwater quality. Monitoring Methodology. EPA-600/4-76-026, Las Vegas, NV

Zojer H, Fank J, Harum T, Leditzky B, Stromberger (1989) Nitratbelastung des Grundwassers im nordöstlichen Leibnitzer Feld. Steir. Beiträge zur Hydogeologie. Ed. FG Joanneum Graz, Styria, Austria, p 7 wastewater from sewer systems, and rainwater. Fig. 4 shows the upward trend in nitrogen use in west German agriculture (Müller, 1990). Disposal of liquid manure on farm fields can be regionally the most important source of ground-water contamination. Rainfall is primarily responsible for nitrate loading in forests (Matzner, 1988; Meiwes and Beese, 1988). In urban areas, increasing nitrate levels are partially due to intensive fertilizer use in gardens, on sporting grounds and on arable land, and also to leaky sewer systems.

If the loading rate (g NO_3^-/m^2 yr) is divided by the yearly ground-water recharge rate (m/yr), the result is the mean concentration of nitrate within the seepage water reaching the ground-water surface. For example, if the loading rate is 20 g NO_3^-/m^2 and the recharge rate is 200 mm/yr = 0.2 m/yr, the NO_3^- concentration is 100 g/m^3 = 100 mg/L. Estimates of nitrate loading rates in rural and urban areas of (former) West Germany are provided in Tables 1 and 2.

Table 1. Approximate loading rates of nitrate in rural ground-water systems in (former) West Germany

Land Use	Fertilization kg N/ha yr	Loading rate [g NO_3^-/m^2 yr] Loamy soil	Sandy soil
Deciduous trees		1 - 2	2 - 4
Conifers		3 - 5	5 - 7
Grassland	up to 150	2 - 4	3 - 5
Arable land: plants cover the	300	4 - 8	7 - 10
surface	150	5 - 7	9 - 11
Arable land: plants cover the	200	12 - 14	15 - 17
surface poorly	150	15 - 17	18 - 21
Areas where liquid manure is disposed (maize)	300	26 - 30	36 - 42
	600	50 - 60	70 - 85

Table 2. Approximate nitrate loading rates in urban ground-water systems in (former) West Germany

Land use	Loading rate [g NO_3^-/m^2 yr]	
City	1 - 2	8 - 10 *
Suburbs	2 - 3	
Garden colonies	10 - 15	
Sporting grounds	3 - 4	

(* if remarkable quantities of sewage water leaves the sewer system)

It is very difficult to estimate the influence of sewage water on the loading of ground water, since this impact depends on the location of the sewer pipes relative to the ground-water surface, and the number and geometry of fissures and cracks in the walls of the pipes. If the pipes are located within the saturated zone, ground water intrudes into the pipes; no sewage water can intrude into the ground water.

4 Output

The major parameters affecting nitrate output from aquifers are denitrification, aquifer geometry, and ground-water velocity. An example of denitrification processes in aquifers is presented in Fig. 5, showing an increase of sulfate in pumping wells near Hannover (Koelle, 1990). Increased levels have also been observed in the Netherlands (Kruithof et al., 1987). Sulfate is produced by the reaction of nitrate with pyrite in the aquifer, assisted by bacteria (Koelle et al., 1985). Denitrification can occur if organic material is available in the aquifer; in fact, organic matter is often introduced around pumping wells as a remedial measure to reduce the nitrate content of pumped water. If no denitrification occurs, the output of nitrate to the pumping wells increases more or less steadily according to the increasing input and decreasing storage capacity of the aquifer. The latter aspect is discussed in the next section.

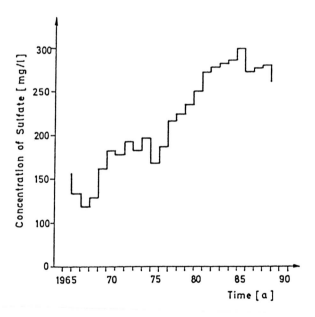

Fig. 5. Increase of sulfate by pyrite oxidation induced by nitrate in ground water, after Koelle (1990)

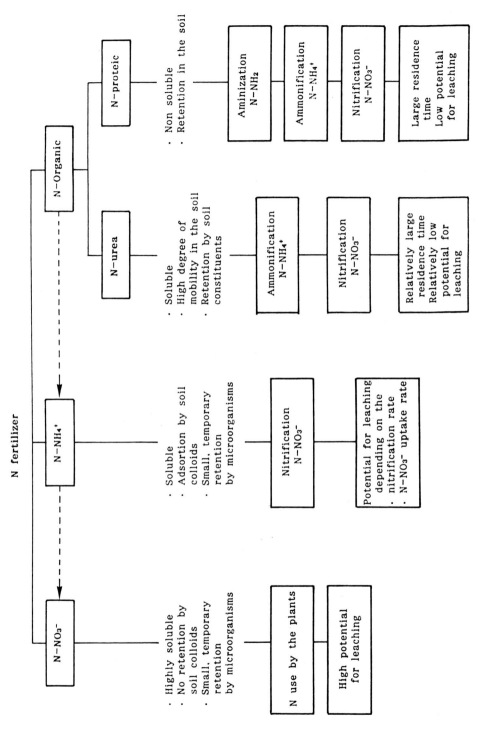

Fig. 1. Potential for nitrate leaching in relation to type of fertilizer

However, the nitrogen contributions from these other sources are difficult to predict because they vary according to the soil chemical profile and farm practices (Power and Broadbent, 1989). Nitrification depends on microbial activity which can be favored by environmental conditions such as the existence of mulch on the soil surface. Mineralization of crop residues and manures is higher when the C/N ratio is smaller, which depends on the mulch characteristics. These sources of variation make an accurate N balance difficult, but nevertheless, the use of accurate fertilizer recommendations definitively reduces excess N in soils.

In order to limit the rate of availability of NO_3-N, nitrification or urease inhibitors may be added to fertilizers containing ammonia or organic N. These inhibitors increase the residence time in the soil and favor a rate of nitrification which is closer to the rate of actual N use by the plants. However, yield responses with nitrification inhibitors are variable, inconsistent and unpredictable (Peterson and Frye, 1989), thus reducing interest in their use. The urease inhibitors (urea in particular) are of particular interest for N-organic fertilizers because they inhibit urease, an enzyme that accelerates the urea hydrolysis, thus stabilizing N in the soil and decreasing urea volatilization. Research is still needed to develop the best inhibitors.

Other fertilization solutions concern the choice of less-soluble fertilizers and specially coated fertilizers like osmocote. These coated fertilizers are produced in a solid form with particles covered by a thin film which "opens" as the temperature increases. As absorption of nutrients by the plants increases with temperature, N becomes readily available when the uptake rate is higher. The leaching of fertilizers may therefore be reduced.

The type of fertilizer also plays an important role for certain crops. In the case of rice in flooded basins, for example, urea or ammonia sulphate are recommended not only because nitrates are more easily lost, but because anaerobic conditions of the soil favor denitrification, and consequently losses of nitrous oxide ($N_{2)}$) and molecular nitrogen (N_2-O).

2.2 Nitrates and Fertilizer Application as Influenced by Agricultural Practices

Methods of fertilizer application, timing, relation of N to other nutrients, cropping systems, and water management also influence N leaching. Fig. 2 summarizes these influences, which combine like a puzzle as analyzed below.

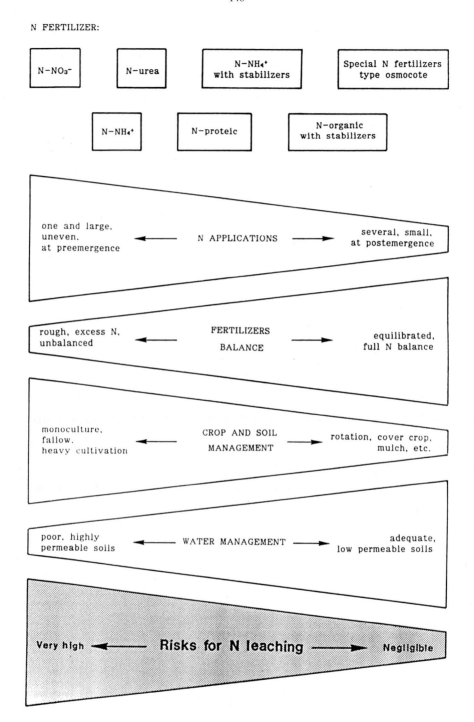

Fig. 2. Risks for nitrate leaching as influenced by type of N fertilizer, fertilizing, and agricultural practices

The method of fertilizer application depends on the type of fertilizer and the time of application. A gaseous fertilizer like anhydrous ammonia has to be injected; a liquid fertilizer may be injected or applied with the irrigation water; a solid fertilizer may be broadcast or banded at the pre-plant or planting stage, but normally should be banded after emergence of the crop (see Tisdale et al., 1985; Peterson and Frye, 1989). Leaching problems arise when application is nonuniform, leading to excessive accumulation of fertilizer in some areas.

The timing of application plays a major role in reducing leaching losses. Large pre-plant or preemergence N applications lead to higher losses because of the delay between application and plant uptake. Postemergence application eases management of N application and improves N efficiency. Split applications timed to immediately precede crop use of N improve fertilizer efficiency, increase yields and reduce leaching potential. However, this is often difficult to achieve, particularly for tall crops. In irrigated crops, fertilizer application with the irrigation water allows timely applications during crop growth, but modernization of on-farm irrigation is required.

The use of N by the plants greatly depends on the availability of other major nutrients such as P, K Ca and Mg, and micronutrients such as Cu, Mo, Mn and Zn. The pH level also may limit nutrient uptake. Consequently, nitrate leaching increases when other nutrients are deficient. Because plant responses to N fertilization are very impressive, farmers tend to forget about these other nutrients which have less obvious effects. In this case, excessive nitrogen is applied, often much more than required, and N losses increase.

Although the primary summer crop plays the major role, several aspects of cropping systems influence nitrate leaching. The existence of a cover crop during winter is of particular importance since the cover crop may significantly reduce nitrate leaching by consuming residual N (Thomas et al., 1989; Magette and Shirmohammadi, 1989; Russelle and Hargrove, 1989; and Juergens-Gschwind, 1989). Intercropping is also likely to favor a reduction of residual N. Several authors point out the negative impact of the crop-fallow system.

Crop rotations which include legumes have been a traditional practice in Mediterranean agriculture, although this has changed recently because prices have favored nonlegume crops. Since the legumes provided a natural source of N, this shift has resulted in increased use of chemical fertilizers. Continued use of legume crops in the rotations is important from the perspective of nitrogen management. When legumes are used in the rotation, the recommended fertilizer rates should be reduced by the amount of N made available in the soil after the legume crop. When the nonlegume crop is irrigated, results are positive if water management provides a good environment for crop development, thus for consuming the residual N (Magette and Shirmohmmadi, 1989; Russelle and Hagrove, 1989). Muller (1989) discusses fertilizer use and the influence of legumes in crop rotation, and presents tables suggesting quantitative approaches for calculating the N balance. This approach is compatible with results on nitrate leaching presented by other authors (Juergens-Gschwind, 1989; Hofman and Verdegem, 1990).

The influence of cultivation techniques is contradictory, and depends highly on soil characteristics, water regime, and crop (Thomas et al., 1989). When cultivation methods favor water movement by loosening the soil and shattering dense, restrictive layers such as plowpans, nitrate leaching can increase because obstacles to water movement are removed and aerobic conditions for nitrification are improved. Comparing conventional and no till treatments, the data show both higher and lower leaching for no-tillage (Thomas et al., 1989; Staver et al., 1989; Maguette and Shirmohmmadi, 1989). Maintenance of surface mulch favors microbial activity because the higher humidity in the top layers increases nitrification (Power and Broadbent, 1989). However, since mulch can also increase infiltration, under certain conditions an anaerobic environment can be favored.

3 Water Management

3.1 General; Modeling Considerations

Aspects analyzed in the previous section relate to excess N in the soil and the potential for leaching losses. However, leaching only occurs when excess water flows downward from the root zone to the ground water; hence, special attention must be given to water management, irrigation and drainage. For irrigated agriculture, the combination of water and fertilizer application deserves particular attention (Feigin, 1987).

Models can be used to evaluate the consequences of the leaching process and the impact of several agricultural practices that influence the hydrologic behavior of soils. Some models concern all terms and processes in the hydrologic balance, like EPIC (Shaffer et al., 1983; Williams et al., 1983). Others focus on the unsaturated zone, like the convective-dispersive flow models (de Smedt and Wierenga, 1979; Biggar and Lu, 1989; Moeller and Veh, 1990). Still others compute the balance of solutes in relation to overland flow (Peters et al., 1979), to irrigation applications and drainage water (Nicholaichuk et al., 1989), or to rainfall and drainage water (Evans et al., 1990; Dvorák, 1990). Stochastic and spatial variability approaches are considered by others (Guitjens, 1990). Interesting reviews on models are presented by Bonazountas (1987) and Jury and Nielsen (1989).

Models help to clarify the processes and explain the variations which occur in field practice. Under certain conditions, they allow one to predict the impacts of given practices (Shaffer et al., 1983). Nevertheless, they may need a site-specific calibration and, in the case of models using unsaturated soil water modules, are sensible to the spatial variability of soil physical characteristics and to the hydrodynamic characteristics of the soil (Jury and Nielsen, 1989). These limitations may impose restrictions on the accuracy of quantitative prediction of solute flow and certainly on the extrapolation of results to other sites or environmental conditions.

Because unsaturated flow is characteristic of the root zone, the unsaturated soil water and solute transport models are of particular interest in evaluating the impacts of irrigation management practices on nitrate leaching. Models used for irrigation management or to simulate water infiltration and redistribution after irrigation utilize the same governing differential flow equations. This compatibility of models is very important because a simulation of irrigation practices can also be used to evaluate drainage or deep percolation losses and, therefore, the potential for solute transport. Examples of such models for irrigation scheduling or irrigation water balance are numerous (Norman and Campbell, 1983; Feyen, 1987; Feddes, 1987; Pereira et al., 1987; de Laat, 1990; Xevi and Feyen, 1990). The use of this modeling approach allows evaluation of the influence of soil hydraulic conductivity on deep percolation and drainage losses with reasonable accuracy (Fernando and Pereira, 1989 and Tabuada, 1989).

Water losses to layers below the root zone are likely to occur in light, sandy soils. In these soils, with very high hydraulic conductivity at saturation and low water-holding capacity, either large volumes or short irrigation intervals often lead to percolation losses, and thus to leaching of nitrates and other nutrients. In contrast, in heavy soils, with low hydraulic conductivity and high water-holding capacity, losses from irrigation only occur when very high irrigation depths are applied. Swelling, cracking soils where preferential flow favors deep water losses are an exception.

3.2 Irrigation

From the perspective of controlling leaching and nitrate pollution (an interesting review has been done by Heermann et al., 1989), two main aspects of irrigation practice are significant: irrigation scheduling and irrigation methods.

Positive environmental impacts of irrigation scheduling programs have been clearly identified (English et al., 1980). In fact, if irrigation depths correspond to the excess soil-moisture deficit, the potential for deep percolation is not drastically reduced. On the other hand, if the timing of water application is such that excessive soil-water deficits are avoided and the crop is not stressed, the uptake of water and nutrients by the plant roots can be maximized, thus reducing the amounts of N remaining in the soil after harvesting. Such assumptions form the basis for irrigation scheduling models which use a sophisticated soil-water balance (Feyen, 1987) as well as for those using simpler approaches but having more general applications (Teixeira, 1989; Teixeira and Pereira, 1990; Smith, 1989; Raes et al., 1988).

Irrigation scheduling under conditions of limited water supply, as in case of drought, is different from scheduling without water restrictions. Fertilizer management also depends on water supply. When water is not a limiting factor, N-fertilization recommendations should aim at full crop requirements. Under drought or water stress conditions, N amounts should be reduced as uptake by the roots will be lower. When fertilization recommendations are not adapted to limited available water or drought conditions, excessive N remains after harvesting, increasing the N-loss potential before the following crop. Stevenson et al. (1986) report an increase of nitrate leaching after a dry year in consequence of lower nutrient uptake by the crop. Heavy nitrate leaching is likely to occur in Mediterranean climates considering the normally heavy fall and winter rains.

Because nitrate is highly mobile, its movement to deeper soil layers occurs as soon as water flows downward. Scheduling irrigation to foster deep root development thus increases the opportunity for nutrient uptake from the lower soil layers. Deep root development is also desirable to make better use of soil water when the soil reservoir has been filled by antecedent rainfall or by a large preemergence irrigation. The first irrigation should then be delayed to induce the roots to explore the entire profile. If N applications precede large preemergent irrigations or rainfall, large N losses may occur. Split fertilizer applications should be used in combination with irrigation scheduling.

Thus adequate irrigation scheduling not only has positive effects on crop growth, but also on the soil N balance, although fertilization strategies must agree with irrigation scheduling decisions. Water management and agricultural practices have to be combined and compatible. The main relationships are shown in Fig. 3.

Irrigation methods play a major role since deep percolation losses depend on the water application process. For surface irrigation, Tabuada (1989) studied several furrow irrigation processes in a highly permeable sandy soil and in a low permeability silt-loam. Irrigation efficiencies for traditional blocked furrows were as low as 36% for the sandy soil but as high as 96% for the silty-loam soil. In the first case, 64% of the water flowed to layers below the root zone, while in the second case, losses were negligible. Hence, this irrigation method should not be used for light soils because of the high potential for water losses and nitrate leaching. However, Fernando and Pereira (1989) found that percolation losses could be reduced to 15% by using better-controlled surface water applications in the same light soil, and Alves (1990) found water losses could be drastically reduced by using sprinkler irrigation. These examples illustrate the need to improve application efficiencies and, often, to use a different method.

In highly permeable soils, sprinkler irrigation (in particular with center pivots or lateral moving systems) can be very appropriate for controlling percolation losses and nutrient leaching. Small but frequent water applications allow farmers to limit the irrigation depth to the desirable allowed soil water deficit; application of fertilizer along with the irrigation water allows regulation of N applications according to crop use, and the uniformity of water and fertilizer distribution can be improved with positive effects on water and solute uptake by the

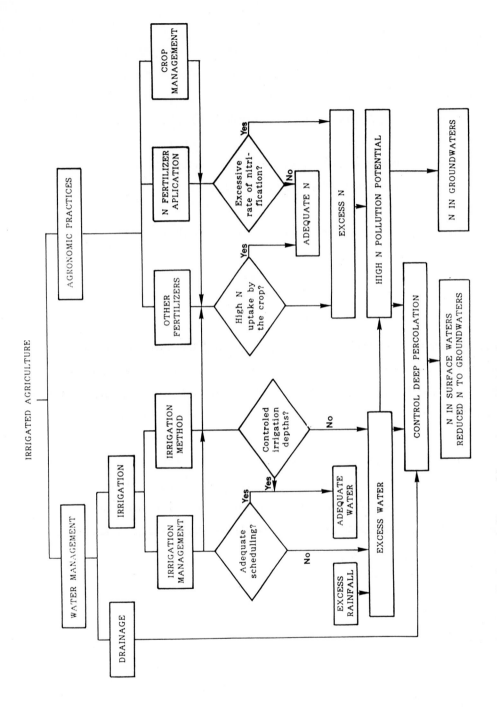

Fig. 3. Nitrate pollution as influenced by water management

crop. Trickle irrigation is a more difficult solution for highly permeable soils because wetted soil volumes tend to be narrow and deep. Both the exigencies for adequate management and the costs increase since more emitters and laterals, and very frequent and light irrigation, are required. Application of nutrients with the irrigation water is also an advantage provided the wetted depths are not deeper than the roots.

For medium-textured soils, all irrigation methods can be applied. However, careful management is still required to achieve high distribution uniformities and application efficiencies.

Heavy soils, with low infiltration rates and high water holding capacity, can also be irrigated by all three methods. The use of center pivots and lateral moving systems is less desirable, however, because infiltration rates can be much lower than application rates, causing runoff and nonuniform water infiltration as well as N losses when the fertilizer is applied with the irrigation water. Trickle irrigation requires careful management to maintain the soil moisture at such levels that soil cracking and consequent water losses are avoided. Surface irrigation methods adapt well to heavy soils, but particular care is required for cracking soils to avoid preferential-flow water losses. Management practices are also a factor in limiting percolation losses from flooded rice fields in soils with low hydraulic conductivity (Pereira, 1989).

Nutrients can also be applied with the irrigation water in surface irrigation, but close supervision is required to minimize losses from runoff and low distribution uniformities. These are two areas of concern in improving on-farm irrigation systems. Level basin irrigation, when fields are laser-leveled, is appropriate for such fertilizer application practices because runoff is controlled and irrigation efficiency and distribution uniformity can be high, providing good control of water and nutrients. Furrow irrigation is another possibility, when the stream size is adequate for the slope, furrow length and infiltration characteristics of the soil. Modern systems of water application to furrows and level basins are required to efficiently apply fertilizers in the irrigation water since better control of irrigation volumes is achieved and irrigation management practices tend to be improved. More research is needed.

The mixture of advantages and disadvantages of different irrigation practices partially explains the controversial results reported in the literature on nitrate leaching. It may be concluded that control of nitrate pollution in irrigated agriculture depends greatly on improvement of irrigation practices, including both irrigation scheduling and irrigation methods. Efforts in these directions will decrease the leaching of N and other nutrients and chemicals, and foster higher yields and better water conservation. In the long run, improvement of irrigation methods in combination with improved fertilizer application techniques promises to limit the impact of nitrate and other chemicals on water quality (Fig. 3).

3.3 Drainage

Drainage practices are used to control the elevation of the water table beneath the crop root zone, thus improving conditions for plant growth (tile or subsurface), or to avoid water-logging problems by diverting the surface waters to open ditches, thus limiting the amount of water penetrating the soil (surface drainage).

In general, the need for drainage controls is limited to fine textured soils with low hydraulic conductivity; that is, soils with the lowest potential for leaching. Nevertheless, because drained lands are under intensive agriculture, the impact of drainage on nitrate pollution deserves particular attention.

As analyzed by Thomas et al. (1989), the effects of surface drainage are mixed: surface drainage limits the excess water infiltrating into the soil that would otherwise cause leaching, but favors aerobic conditions which increase nitrification and decrease denitrification losses of N in a volatile form. Nevertheless, the advantages may outweigh the disadvantages since the environment for the crop root system becomes more favorable for using the available nutrients.

Subsurface drainage is similarly controversial, but here again the advantages prevail: nitrates are intercepted with the drainage water, decreasing the amount of leachate that would otherwise percolate to the water table. If disposal of drainage waters is adequate, then environmental benefits can be added to the agricultural and economic ones. These assumptions are compatible with those relative to irrigation return flow management (Skogerboe et al., 1979) and agree with recent research results discussed below.

Richard et al. (1989) report that drainage altered the redox status of the soil, increasing nitrate loss in the drained effluent, in contrast to the positive changes which occurred in the aerobic condition of the soil. Nevertheless, the potential for such modification is limited, and Elder and Chieng (1989) found that while nitrate concentration decreased with drain depth, it did not depend on the water-table elevation. The authors attributed this to a removal of nitrogen from the percolating solution by either denitrification or immobilization. Negative impacts of drainage on decreasing denitrification are somewhat limited and depend on prevailing local conditions.

Nitrate concentrations in drainage water increase with hydraulic conductivity (Elder and Chieng, 1989; Katalin, 1990; Evans et al., 1990). This increase is related both to the drainage flow rate and to the amount of denitrification, because air-filled porosity is generally lower in less permeable soils which are associated with more favorable conditions for anaerobic processes. The dependence of nitrate leaching on hydraulic conductivity tends to increase with drainage efficiency and drainage intensity (Evans et al., 1990), although this is disputed (Arlot, 1990). Differences are related to local soil conditions, hydrologic regime, crop system and drainage facilities. Other field observations also agree with the above considerations and indicate that variations in nitrate concentration are related to the time and rate of fertilizer application, and amount and distribution of precipitation (Monke et al., 1989; Katalin, 1990).

High nitrate levels in subsurface drainage waters may require special solutions to control disposal in catchments where domestic water-supply facilities are located. Such solutions include retardation or control drainage systems for decreasing drainage discharges, and special drainage disposal systems and reservoirs for treatment of drainage effluents (Dvorák, 1990). Under certain conditions, surface drainage may be considered instead of tile drainage.

In conclusion, surface drainage reduces excess soil water and thus the potential for N leaching, while tile drainage intercepts water and solutes which otherwise would flow downwards to the ground water.

Fig. 3 summarizes the main influences of water management practices on nitrate leaching and how these relate to different agricultural practices. It becomes clear that neither fertilizer

has to combine training, demonstration and extension activities with incentives and policy measures favoring the adoption of improved techniques by farmers. A monitoring activity is desirable.

Many potential solutions are well-known, their effectiveness has been proven, and the related technologies are available. Nevertheless, they have not yet become common practice become adoption of innovations occurs slowly unless the economic benefits are clear. Therefore, development and implementation of best management practices to control nitrate pollution is not only a matter of scientific and technological development but a subject of environmental and agricultural policy. This multidimensional approach can give coherence to a program and ensure compatibility among scientists, extensionists and farmers.

References

Alves IM (1990) Water-yield modelling approaches, application to maize (in Portuguese). M.Sc. thesis, Technical University of Lisbon, Lisbon

Arlot MP (1990) Exportations d'azote par les eaux de drainage souterrain. In: Holy M et al. (eds) The influence of irrigation and drainage on the environment with particular emphasis on impact on the quality of surface and ground waters (Trans 14th ICID Congress). ICID, New Delhi, p 345

Bashkin V (1989) Accumulation of agrochemicals and their metabolites in surface and ground water of various agroclimatic regions. In: Summers JB, Anderson SS (eds) Toxic substances in agricultural water supply and drainage, an international environmental perspective. US Committee on Irrigation and Drainage, Denver, p 367

Biggar JW, Lu TX (1989) Modelling soil-water processes to evaluate environmental pollution. In: Dodd VA, Grace PM (eds) Agricultural engineering. 1. Land and water use (Proc. 11th Int. Cong. CIGR). Balkema, Rotterdam/Brookfield, p 255

Bonazountas M (1987) Chemical fate modelling in soil systems: a state-of-the-art review. In: Barth H, L'Hermite P (eds) Scientific basis for soil protection in the European Community. Elsevier, London, p 487

Bower H (1987) Effect of irrigated agriculture on groundwater. J Irrig Drain Engin (ASCE), 113(1), p 4

De Haan FAM (1987) Effects of agricultural practices on the physical, chemical and biological properties of soils, Part III. Chemical degradation of soil as the result of use of mineral fertilizers and pesticides: aspects of soil quality evaluation. In: Barth H, L'Hermite P (eds) Scientific basis for soil protection in the European Community. Elsevier, London, p 211

De Laat PJM (in press) MUST, a pseudo steady-state approach to simulating flow in unsaturated media. In: Workshop on Crop and Water Models, ICID, Rio de Janeiro, April 1990

De Smedt F, Wierenga PJ (1979) Simulation of water and solute transport in unsaturated soils. In: Morel-Seytoux, HJ et al. (eds) Surface and subsurface hydrology. Water Resources Publications, Fort Collins, p 430

Dodd VA, Grace PM (eds) (1989) Agricultural engineering. 1. Land and water use (Proc. 11th Congress CIGR, Dublin, September). Balkema, Rotterdam

Dvorák P (1990) Impact of drainage on the quality of surface waters. In: Holy M et al. (eds) The influence of irrigation and drainage on the environment with particular emphasis on impact on the quality of surface and ground waters (Trans. 14th ICID Congress). ICID, New Delhi, p 263

Elder L, Chieng ST (1989) Effect of water table height and physical properties on nutrient leaching. In: Summers JB, Anderson SS (eds) Toxic substances in agricultural water supply and drainage, an international environmental perspective. US Committee on Irrigation and Drainage, Denver, p 293

English MJ, Horner GL, Orlob GT, Erpenbeck J, Moehlman M, Cuenca RH, Dudek DJ (1980) A regional assessment of the economic and environmental benefits of an irrigation scheduling service. US-EPA, Ada. Oklahoma

Evans RO, Gillian JW, Skaggs RW (1990) Controlled drainage and subirrigation effects on drainage water quality. In: Holy M et al. (eds) The influence of irrigation and drainage on the environment, with particular emphasis on impact on the quality of surface and ground waters (Trans. 14th Congress of ICID). ICID, New Delhi, p 13

Feddes RA (1987) Modelling and simulation in hydrologic systems related to agricultural development: state of the art. In: Feyen J (ed) Simulation models for cropping systems in relation to water management. Com. European Communities, EUR 10869, Luxembourg, p 55

Feigin A (1987) Strategies for soil protection under intensive irrigation in Israel. In: Barth H, L'Hermite P (eds) Scientific basis for soil protection in the European Community. Elsevier, London, p 471

Fernando RM, Pereira LS (1989) Evaluation of irrigation management practices for corn in a sandy soil and in a loamy soil. In: Plancquaert P (ed) Management of waters resources in cash crops and in alternative production systems. Com. European Communities, EUR 11935, Luxembourg, p 53

Feyen J (1987) Field validation of soil water and crop models. In: Feyen J (ed) Simulation models for cropping systems in relation to water management. Com. European Communities, EUR 10869, Luxembourg, p 105

Follet RF (ed) (1989) Nitrogen management and ground water protection. Elsevier, Amsterdam

Follet RF, Walker DJ (1989) Groundwater quality concerns about nitrogen. In: Follet RF (ed) Nitrogen management and ground water protection. Elsevier, Amsterdam, p 1

Ghaly AE (1989) Air and ground water pollution from high application of animal manure. In: Dodd VA, Grace PM (eds) Agricultural engineering. 1. Land and water use. (Proc. 11th Int. Cong. CIGR). Balkema, Rotterdam/Brookfield, p 415

Guitjens JC (1990) Nutrient estimation of variable drainage. In: Holy M et al. (eds) The influence of irrigation and drainage on the environment, with particular emphasis on impact on the quality of surface and groundwaters (Trans. 14th ICID Congress). ICID, New Delhi, p 65

Heermann DF, Duke HR, van Schilfgaarde J (1989) Management of water balance components. In: Follet RF (ed) Nitrogen management and ground water protection. Elsevier, Amsterdam, p 319

Hofman G, Verdegem L (1990) Nitrate migration and drainage losses in arable cropping systems under a temperate marine climate. In: Holy M et al. (eds) The influence of irrigation and drainage on the environment with particular emphasis on impact on the quality of surface and ground waters (Trans. 14th ICID Congress). ICID, New Delhi, p 221

Holy M et al. (ed) (1990) The influence of irrigation and drainage on the environment with particular emphasis on impact on the quality of surface and ground waters (Trans. 14th Congress on Irrigation and Drainage, Rio de Janeiro, April) vol I-B and C, ICID, New Delhi

Juergens-Gschwind S (1989) Ground water nitrates in other developed countries (Europe): Relationships to land use patterns. In: Follet RF (ed) Nitrogen management and ground water protection. Elsevier, Amsterdam, p 75

Jury WA, Nielsen DR (1989) Nitrate transport and leaching mechanisms. In: Follet RF (ed) Nitrogen management and ground water protection. Elsevier, Amsterdam, p 139-157

Katalin KH (1990) Nitrogen migration in soils and water. In: Holy M et al. (ed) The influence of irrigation and drainage on the environment, with particular emphasis on impact on the quality of surface and ground waters (Trans. 14th ICID Congress). ICID, New Delhi, p 123

Magette WL, Shirmohammadi A (1989) Nitrate in shallow unconfined ground water beneath agricultural fields. In: Dodd VA, Grace PM (eds) Agricultural engineering. 1. Land and water use (Proc. 11th Int. Cong. CIGR). Balkema, Rotterdam/Brookfield, 297

Moeller W, Veh GM (1990) The influence of irrigation on the groundwater recharge and the leaching of nitrate. In: Holy M et al. (eds) The influence of irrigation and drainage on the environment, with particular emphasis on impact on the quality of surface and ground waters (Trans. 14th ICID Congress). ICID, New Delhi, p 207

Monke EJ, Kladivko EJ, Van Scoyoc GE, Martinez MA, Huffman RL (1989) Movement of pesticides and nutrients in drainage water. In: Dodd VA, Grace PM (eds) Agricultural engineering. 1. Land and water use (Proc. 11th Int. Cong. CIGR). Balkema, Rotterdam/Brookfield, p 305

Muller JC (1989) Un outil pour la maîtrise de l'azote en agriculture: la méthode du bilan. In: Dodd VA, Grace PM (eds) Agricultural engineering. 1. Land and water use (Proc. 11th Int. Cong. CIGR). Balkema, Rotterdam/Brookfield, p 519

Nicholaichuk W, Whiting J, Grover R (1989) Herbicide, nutrient and water loss when irrigating by the corrugation method. In: Summers JB, Anderson SS (eds) Toxic substances in agricultural water supply and drainage, an international environment perspective. US Committee on Irrigation and Drainage, Denver, p 269

Norman JM, Campbell G (1983) Application of a plant-environment model to problems in irrigation. In: Hillel D (ed) Advances in irrigation, vol II. Academic Press, New York, p 155

Parton WJ, Persson J, Anderson DW (1983) Simulation of organic matter changes in swedish soils. In: Lauenroth WK, Skogerboe GV, Flug M (ed) Analysis of ecological systems: state-of-the-art in ecological modelling. Elsevier, Amsterdam, p 511

Pereira LA (1989) Management of rice irrigation (in Portuguese). Ph.D. thesis, Instituto Superior de Agronomia, Lisbon

Pereira LS, Teixeira JL, Pereira LA, Ferreira MI, Fernando RM (1987) Simulation models of crop responses to irrigation management: research approaches and needs. In: Feyen J (ed) Simulation models for cropping systems in relation to water management. Com. European Communities, EUR 10869, Luxembourg, p 19

Peters RE, Lee CR, Bates DJ, Reed BE (1979) Influence of storms on nutrient runoff from an overland flow land treatment system. In: Proc. Hydrologic Transport Modelling Symp.. ASAE, St. Joseph, Michigan, p 162

Peterson GA, Frye WW (1989) Fertilizer nitrogen management. In: Follet RF (ed) Nitrogen management and ground water protection. Elsevier, Amsterdam, p 183

Power JF, Broadbent FE (1989) Proper accounting for N in cropping systems. In: Follet RF (ed) Nitrogen management and ground water protection. Elsevier, Amsterdam, p 159

Raes D, Lemmers H, Van Aelst P, Vanden Bulcke M, Smith M (1988) IRSIS-Irrigation scheduling information systems. Lab. of Land Management, K.U. Leuven, Leuven

Richard P, Nagpal NK, Chieng ST (1989) Water and nutrient movement in an agricultural soil under drained and undrained conditions. In: Summers JB, Anderson SS (ed) Toxic substances in agricultural water supply and drainage, an international environmental perspective. US Committee on Irrigation and Drainage, Denver, p 281

Russelle MP, Hargrove WL (1989) Cropping systems: ecology and management. In: Follet RF (ed) Nitrogen management and ground water protection. Elsevier, Amsterdam, p 277

Santos JQ (1986) Fertilization and soil pollution (in Portuguese). Pedon (5):85-97

Santos JQ (1988) Soil fertility and fertilization (in Portuguese). In: 1st Symposium on Desertification, Universidade de Évora, Évora

Santos JQ (1989) Reflexions sur la pollution chimique des sols au Portugal: action des fertilizants. In: Rèunion sur surveillance de l'état de santé des sols dans le cadre du program CORINE, EEC, Bruxelles

Santos JQ (1990) Influence of fertilizers on agricultural progress (in Portuguese). In: 6th Congress of Algarve, Racal Clube, Albufeira, vol II, p 531

Santos JQ, Vasconcelos E, Cabral F, Monjardino P (1988) Pollutant organic wastes as fertilizers. In: Storing, handling and spreading of manure and municipal waste (CIGR Seminar, Uppsala, September). Swedish Institute of Agriculture Engineering, Uppsala, p 4.1

Sauerbeck D (1987) Effects of agricultural practices on the physical, chemical and biological properties of soils: Part II-Use of sewage sludge and agricultural wastes. In: Barth H, L'Hermite P (ed) Scientific basis for soil protection in the European Community. Elsevier, London, p 181

Schmidt KD, Sherman I (1987) Effect of irrigation on groundwater quality in California. J Irrig Drain Engin (ASCE), 113(1):16-29

Shaffer MJ, Gupta SC, Linden DR, Molina JAE, Clapp CE, Larson WE (1983) Simulation of nitrogen, tillage and residue management effects on soil fertility. In: Lauenroth WK, Skogerboe GV, Flug M (eds) Analysis of ecological systems: state-of-the-art in ecological modelling. Elsevier, Amsterdam, p 525

Skogerboe GV, Walker WR, Evans RG (1979) Environmental planning manual for salinity management in irrigated agriculture. US-EPA, Ada, Oklahoma

Smith M (1989) Manual for CROPWAT, a computer program for IBM-PC or compatibles. Land and Water Division, FAO, Rome

Staver K, Stevenson JC, Brinsfield R (1989) The effect of best management practices on nitrogen transport in to Chesapeake Bay. In: Summers JB, Anderson SS (eds) Toxic substances in agricultural water supply and drainage, an international environmental perspective. US Committee on Irrigation and Drainage, Denver, p 163

Stevenson JC, Brinsfield R, Staver K (1986) Surface runoff and groundwater impacts from agricultural activities in the Chesapeake region. In: Summers JB, Anderson SS (ed) Toxic substances in agricultural water supply and drainage, defining the problem. US Committee on Irrigation and Drainage, Denver, p 211

Summers JB, Anderson SS (ed) (1989) Toxic substances in agricultural water supply and drainage, an international environmental perspective. US Committee on Irrigation and Drainage, Denver

Tabuada MA (1989) Two- and three-dimensional modelling of furrow irrigation (in Portuguese). Ph.D. thesis, Instituto Superior de Agronomia, Lisbon

Teixeira JL (1989) Modelling for planning and scheduling irrigations (in Portuguese). Ph.D. thesis, Instituto Superior de Agronomia, Lisbon

Teixeira JL, Pereira LS (in press) ISAREG, an irrigation scheduling simulation model. In: Workshop on Crop Water Models, ICID, Rio de Janeiro, April 1990

Thomas GW, Smith MS, Phillips RE (1989) Impact of soil management practices on nitrogen leaching. In: Follet RF (ed) Nitrogen management and ground water protection. Elsevier, Amsterdam, p 247

Tisdale SL, Nelson WL, Beaton JD (1985) Soil fertility and fertilizers (4th ed.). Macmillan Publishers Co., New York

Williams JR, Dyke PT, Jones CA (1983) EPIC-a model for assessing the effects of erosion on soil productivity. In: Lauenroth WK, Skogerboe GV, Flug M (eds) Analysis of ecological systems: state-of-the-art in ecological modelling. Elsevier, Amsterdam, p 553

Xevi E, Feyen J (in press) Combined soil water dynamic model (SWATRER) and summary crop simulation model (SUCROS). In: Workshop on Crop Water Models, ICID, Rio de Janeiro, April 1990

INTEGRATED WATER AND NITROGEN MANAGEMENT

J.S. Schepers, D.L. Martin[1], D.G. Watts,[2] and R.B. Ferguson[3]
Soil Scientist, USDA-Agricultural Research Service
Agronomy Department, 113 Keim Hall
University of Nebraska-Lincoln
Lincoln, Nebraska 68583-0915 U.S.A.

Abstract

Nitrate contamination of ground water is a concern throughout the world. Control of nitrate leaching is necessary to protect or improve water quality, but effective management is difficult because of the complex interactions between soil, water, and nitrogen (N). It is widely recognized that N and water management must be addressed simultaneously to develop production systems that reduce nitrate leaching. Difficulties with N and water management arise because of the various sources of N, uncertainties in N cycling and nitrate leaching imposed by climatic variability, and logistic considerations on the part of producers. Traditional approaches to N management usually involve applying ample fertilizer N early in the growing season to meet anticipated crop needs. An alternative is to apply small amounts of fertilizer early in the growing season and use tissue analysis to schedule additional fertilizer N as the season progresses. Fertilizer N can be applied in the irrigation water when the crop is too tall for typical machinery. Both leaf N concentration and use of chlorophyll meters are promising techniques for evaluating the N status of irrigated corn. Crop N and water needs can also be estimated using simulation models. Computer simulations and tissue testing procedures combined with improved irrigation techniques can help develop farming systems that conserve both water and N, while maintaining yields and reducing nitrate leaching.

1 Introduction

Producer attitudes about N and water management for profitable production are difficult to change. While producers desire to protect and preserve ground-water resources, there is little

---

[1] Biological Systems Engineering, 231 Chase Hall, University of Nebraska-Lincoln, Lincoln, NE 68583-0726 U.S.A.
[2] Biological Systems Engineering, 230 Chase Hall, University of Nebraska-Lincoln, Lincoln, NE 68583-0726 U.S.A.
[3] South Central Research and Extension Center, Clay Center, NE 68933, U.S.A.

NATO ASI Series, Vol. G 30
Nitrate Contamination
Edited by I. Bogárdi and R. D. Kuzelka
© Springer-Verlag Berlin Heidelberg 1991

and one third of the 970 bores sampled in this area have concentrations exceeding 10 mg NO_3-N/L (Dillon, 1988). Although most of this nitrate is derived from leaching beneath leguminous pastures which are grazed by livestock (Dillon, 1989a, 1989b), the highest concentrations are due to point sources, such as wastes from saleyards, dairies, piggeries, and meat and milk processing. As ground water is the sole source of piped water supply for the City of Mount Gambier and surrounding townships, a better understanding of the fate of relic nitrate plumes from point sources was demanded for effective water resources management.

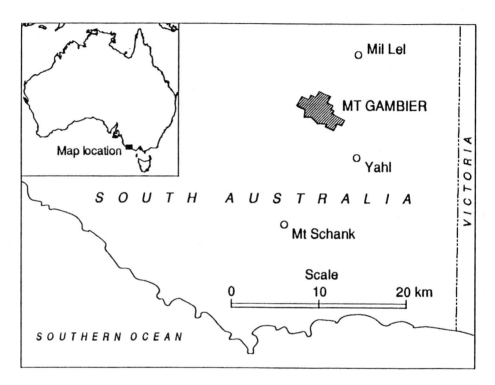

Fig. 1. Location of Yahl study site

2 Site Description

Yahl is located on the Gambier Plain consisting locally of solodic soils to a depth of 1 m overlying carbonaceous clay to 4 m then Gambier limestone which is about 150 m thick in this vicinity. This Tertiary karstic limestone is an unconfined regional aquifer and the most significant source of ground water used in the area. The depth to water table is 20 m, and unlike the regional water table which slopes to the south, the local gradient is towards the northwest. Grazing of beef, sheep and dairy cattle is the main land use and some grain, fodder and vegetable crops are also grown.

The area has cool wet winters and warm dry summers with an annual rainfall of about 800 mm and pan evaporation of about 1500 mm. Rainfall exceeds pasture water use between April and September (Smith and Schrale, 1982), and the annual recharge has been determined to be 150 mm in this area (Allison and Hughes, 1978). Mean monthly temperatures vary seasonally between 9 and 18°C.

3 Plume Identification

A series of surface geophysical surveys were conducted to determine the extent of ground-water contamination from effluent disposal at Yahl. Electrical Resistivity (ER), Frequency Domain Electromagnetics (FEM), and Time Domain Electromagnetics (TEM) were used to map changes in electrical conductivity in the upper part of the saturated aquifer. The effluent had a relatively high electrical conductivity (200-700 mS/m) compared with the ambient ground water (50-100 mS/m). This enabled significant differences in ground-water conductivity to be detected (Fig. 2).

Results were consistent for the three methods and enabled drilling targets to be selected. Richardson (1990) describes these techniques and results in detail for Yahl as well as two other sites in the region. Six holes were drilled to at least 10 m below the water table at target sites on two transects, one along the "axis" of the plume and the other perpendicular. Electrical conductivities of ground-water samples were in agreement with those predicted from the geophysical surveys.

In the limestone the holes were uncased and ground water was sampled at selected depths using an electric submersible pump between two inflatable packers. Chemical and bacterial sampling and analysis procedures are described by Richardson (1990).

4 Plume Composition

The plume was found to consist of a regular chemical structure, and microbial species and their numbers gave a coherent view of the processes occurring in the plume. Three zones were found with distinct chemical characteristics, showing nitrogen and carbon species in concentrations of oxygen and low concentrations of organic carbon are limiting denitrification. Therefore, the nitrate produced in the plume will not degrade without intervention.

Although no pollutants have been injected into the aquifer for thirteen years there is still potential for the formation of nitrate. This may be of concern for water supply considering the large number of relic point-source plumes in the vicinity, of which Yahl is one example. A possible remedial strategy is to inject oxygen or an oxidizing agent into Zone 1 to allow conversion of nitrogen compounds in reduced form to nitrate. Once the conversion was complete (such as occurs in Zone 3), a cheap carbon source could be used to induce denitrification and so clean up the source of contamination.

Acknowledgements

This study was supported by the South Australian Department of Mines and Energy, The Engineering and Water Supply Department of South Australia, a Partnership Research Grant (P88/10) of the Australian Water Research Advisory Council, and a postgraduate scholarship of the Centre for Groundwater Studies. The donation of packers by AGE Developments Pty. Ltd., Perth, Australia, and use of the South Australian Department of Agriculture microbiology laboratory in Mount Gambier is gratefully acknowledged. Field assistance was provided by P.Cook, J. Dighton, A. Holub and R. Scott. Dr. A. Herczeg and Dr. C. Barber CSIRO Division of Water Resources assisted through helpful discussions.

References

Allison GB, Hughes MW (1978) The use of environmental chloride and tritium to estimate total recharge to an unconfined aquifer. Aust J Soil Res 16:181-195

Baedecker MJ, Back W (1979) Hydrogeological processes and chemical reactions at a landfill. Ground Water 17:429-437

Dillon PJ (1988) An evaluation of the sources of nitrate in groundwater near Mount Gambier, South Australia. CSIRO Water Resources Series No.1

Dillon PJ (1989a) An analytical model of contaminant transport from diffuse sources in saturated porous media. Water Resources Research 25:1208-1218

Dillon PJ (1989b) Models of nitrate transport at different space and time scales for groundwater quality management. In: Jousma G et al. (eds) Groundwater contamination: use of models in decision making. Kluwer Academic Publishers, Dordrecht, p 273

Richardson SB (1990) Groundwater contamination by cheese factory and abattoir effluent. Flinders University of South Aust. School of Earth Sciences, MSc Thesis

Smith PC, Schrale G (1982) Proposed rehabilitation of an aquifer contaminated with cheese factory wastes. Aust Water and Wastewater Assoc J Water 9:21-24

Waterhouse JD (1977) The hydrogeology of the Mount Gambier area. Report of investigations No.48. Dept. of Mines. Geological Survey of South Australia

NITRATE CONTAMINATION OF GROUND WATER IN SOUTHERN ONTARIO AND THE EVIDENCE FOR DENITRIFICATION

R.W. Gillham
Waterloo Centre for Groundwater Research
University of Waterloo
Waterloo, Ontario
N2L 3G1 Canada

Abstract

There has been no comprehensive survey of nitrate concentrations in aquifers of southern Ontario. Nevertheless, the results of several site-specific studies suggest that nitrate contamination is ubiquitous in surficial aquifers in areas of intensive agricultural activity, with concentrations commonly exceeding the drinking water limit of 10 mg/L NO_3-N. There is little doubt that agricultural fertilizer is a major contributor to the elevated nitrate values. Based on several lines of evidence, including vertical profiles of nitrate and dissolved oxygen concentration, it appears that in some aquifers denitrification is an active process mitigating against widespread nitrate contamination. The occurrence of denitrification appears to be controlled by the transport of labile organic carbon from the soil zone to the ground-water zone. Enhanced carbon transport could promote denitrification, resulting in improved ground-water quality, and the potential for denitrification is a factor that should be considered in delineating nitrate-sensitive land areas. In order to exploit the potential benefits of denitrification in aquifers, an improved understanding of carbon cycling in the unsaturated zone is required.

1 Introduction

The Province of Ontario, in Canada, has an area of approximately 1 million km^2 and a population of 8.6 million people. While the low population density suggests that anthropogenic effects on the environment should be small, it is important to recognize that about 90% of the population is concentrated on 15% of the land area. That is, about 7.5 million people occupy an area of land about 150 km wide bordering on the Great Lakes. Fig. 1 includes the area of highest population density. In addition to being the most populated area of the province, this is the most industrialized and, for both geologic and climatic reasons, also the area of most intensive agricultural production in the province.

NATO ASI Series, Vol. G 30
Nitrate Contamination
Edited by I. Bogárdi and R. D. Kuzelka
© Springer-Verlag Berlin Heidelberg 1991

Nitrate concentrations ranged from 0.0 to 70 mg/L NO_3-N, with considerable variation in both the horizontal and vertical directions. Considering only the upper part of the aquifer (to a depth of about 2 m below the water table) nitrate concentrations exceeded 5 mg/L NO_3-N over most of the study area and, over about one-third of the area, concentrations exceeded the drinking water limit of 10 mg/L NO_3-N. Low concentrations (less than 1.0 mg/L) were restricted to wooded areas or other areas of low agricultural activity. It is clear that a significant portion of the aquifer has become contaminated by nitrate and, from a consideration of possible sources, there is little doubt that the contamination has occurred as a result of nitrate being leached from the soil surface.

Fig. 2 is a vertical cross section through the study area showing the geology, topography, and the generalized direction of ground-water flow. Of particular note are the nitrate and dissolved oxygen concentrations determined at the sampling points along the cross section. Elevated nitrate concentrations were limited to shallow depths below the water table (generally one to two meters), while at greater depths, concentrations were consistently less than 1 mg/L NO_3-N. The division between the high- and low-nitrate zones was remarkably sharp, with concentrations declining by as much as 10 mg/L over a vertical distance of 1 m or less. The vertical trend in dissolved oxygen (D.O.) concentration is similar to that of nitrate. In particular, the high nitrate concentrations at shallow depth correspond with high D.O. values, generally between 2 and 10 mg/L, while the sharp decline in nitrate concentration corresponds with the region in which D.O. values show a generally sharp decline to values less than about 2 mg/L. Though chloride concentrations (not shown) were found to be variable, spatial trends were not evident. Assuming agricultural fertilizer to be the main source of both chloride and nitrate, and considering chloride to be a nonreactive solute, the dissimilarity in the spatial distribution of nitrate and chloride was taken as evidence that nitrate was being lost through chemical processes.

Based on the trends in nitrate and dissolved oxygen concentrations (Fig. 2), the dissimilarities in the chloride and nitrate distributions, the physical hydrogeology and the distributions of tritium, oxygen-18 and deuterium, Hendry et al. (1983) concluded that the vertical decline in nitrate concentration was a consequence of the process of denitrification. Because the trends were consistent over the entire study area and over the two-year monitoring period, it is further reasonable to conclude that, at this particular site, denitrification is an important

185

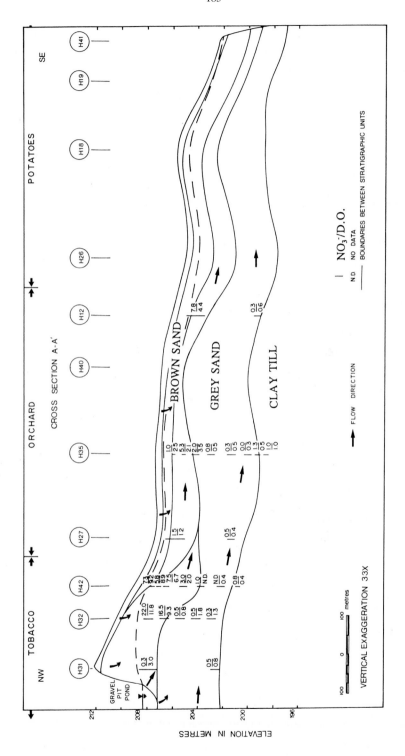

Fig. 2. Cross section through the Hillman Creek study area (NO₃⁻/D.O.)

and persistent process over significant areas and over time. It is also important to note that without vertical profiles of sampling points, the vertical trend in geochemical parameters would not have been observed and thus the apparent significance of denitrification would not have been recognized.

2.2 Alliston

The surficial aquifer in the Alliston region (approximately 80 km north-northwest of Toronto) consists of fluvial sand deposits in a broad, generally flat area, adjacent to the Nottewasaga River. The aquifer is confined laterally on both sides of the river by till outcrops and varies in thickness from a few meters to several tens of meters. The surficial aquifer is separated from a lower confined aquifer by a thick clayey till aquitard. The lower aquifer, which does not have elevated nitrate concentrations, is the main source of water for the town of Alliston; however, private wells in the region generally consist of sand points driven to relatively shallow depths in the surficial aquifer. The area is under intensive agricultural production, with potatoes as the primary crop.

Hill (1982) reported the results of a survey of nitrate concentrations in water samples collected from domestic wells in the area. Concentrations ranged from 0.0 to almost 100 mg/L NO_3-N. The distribution of high nitrate values is shown in Fig. 3. Of particular note, the boundaries of the high-nitrate region generally fall within the boundaries of intensive agricultural production. From an analysis of land use and rates of fertilizer application, it was concluded that agricultural fertilizer was a major source of nitrate to the ground-water zone. From the consistency of the occurrence of nitrate contamination, and from a consideration of nitrate and chloride budgets, Hill (1986) concluded that denitrification was not a significant sink for nitrate in the surficial aquifer at Alliston.

To provide a basis for comparison with the results obtained at Hillman Creek, the University of Waterloo installed several nests of sampling wells in the Alliston aquifer. A sample of the results is given in the cross section of Fig. 4. The results were consistent with those of Hill (1982) in that substantially elevated nitrate values were found where the aquifer was overlain by cultivated areas. As reflected in Fig. 4, however, there was considerable variation in nitrate concentration with depth. At the right-hand side of the cross section where the water

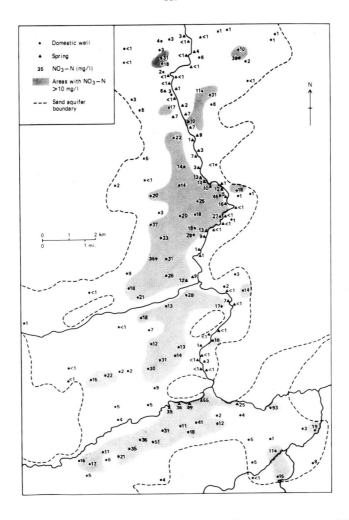

Fig. 3. Map of the surficial sand aquifer at Alliston, showing areas of high nitrate concentration (from Hill, 1982)

table was reasonably shallow (sampling locations MC-1,3,5), nitrate concentrations at depth were consistently less than 1 mg/L NO$_3$-N; while at shallow depths, the concentrations were high (shallow data was not available for MC-1). In this same region, the chloride concentrations were variable but did not show a consistent spatial trend, and the D.O. values, available only for MC-3, showed a decline in concentration with depth. Though certainly less convincing than the data from Hillman Creek, the vertical trends at MC-1, 3, and 5 suggest that denitrification may indeed be occurring in the vicinity of those sampling locations. In the region where the water table was at significant depth (MC-6,7), high nitrate concentrations were encountered at all sampling depths. Though D.O. values were not determined at these particular sampling locations, at similar locations within the basin (deep water table conditions), D.O. values were greater than 3 mg/L. Thus, at these locations, denitrification did not appear to be an active process.

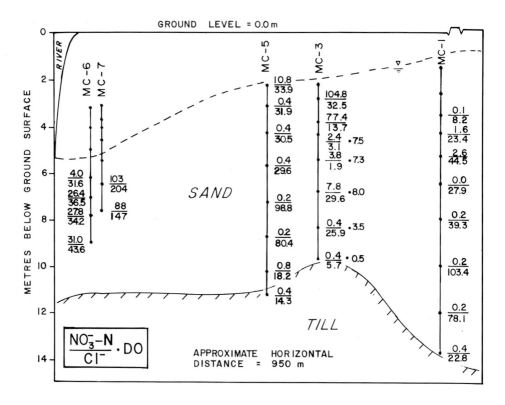

Fig. 4. Cross section through a portion of the Alliston aquifer showing NO_3^-, Cl, and D.O. concentrations

Though the geologic and land-use characteristics of the Hillman Creek and Alliston aquifers are quite similar, the distribution and apparent behavior of nitrate within the aquifers is quite different. Specifically, while the Hillman Creek aquifer shows consistent evidence of denitrification, the Alliston aquifer shows evidence that denitrification may be occurring in some areas; while in other areas, there is strong evidence that denitrification is not a significant process. It is also evident that because of the high degree of vertical variability in nitrate concentrations in both aquifers, samples collected from domestic wells provide a poor basis for evaluating the transport behavior of nitrate in the aquifers.

2.3 Nitrate-Dissolved Oxygen Survey

The data obtained at Hillman Creek and the more recent data from the Alliston aquifer suggest a strong relationship between the occurrence of nitrate contamination and the presence of dissolved oxygen. A survey of nitrate and dissolved oxygen concentrations in eight areas of southern Ontario, having a range in land use and hydrogeologic characteristics, was reported by Egboka (1978). In addition, more detailed site investigations in which nitrate and dissolved oxygen were measured are reported in Hendry (1977), Stiebel (1977), Trudell et al. (1986), Gillham et al. (1984) and Starr and Gillham (1989). In total, twelve areas have been examined; the locations and general land use and geologic conditions of each site are given in Fig. 1.

The sites can be grouped into three broad categories:
1. Surficial sandy aquifers with intensive agricultural production
2. Surficial sandy aquifers with little or no agricultural production
3. Fine-textured surficial materials with or without intensive agricultural production

The sites were not selected systematically, and often included sites to which we had access through other projects. As a result, each land use/geologic category does not have equal representation in the data. Indeed, close to 80% of the data is from surficial sand aquifers. It should be noted, however, that all samples were collected from installations of multilevel sampling wells rather than from domestic supply wells.

The database, including the survey and supplementary data from site-specific studies, now includes over 800 paired nitrate and dissolved oxygen values. Though worthy of a more detailed discussion, only the most general trends in the data, as represented in Fig. 5, will be presented here.

Area A of Fig. 5 represents samples that had a NO_3-N concentration of less than 1.5 mg/L and a dissolved oxygen concentration of less than 2 mg/L. Almost 50% of the data points fall within this region, including most of the data points from both cultivated and uncultivated fine-textured sites. Typical of the data from Hillman Creek, many, though certainly not all, of the points from sandy cultivated regions also fall within area A.

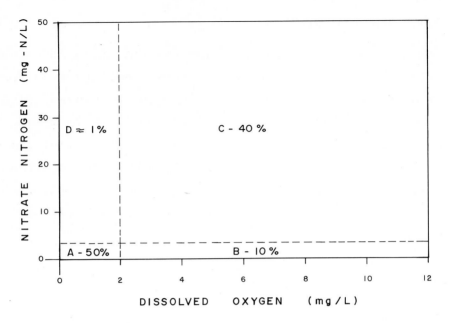

Fig. 5. Distribution of nitrate and dissolved oxygen data measured at the study areas indicated in Fig. 1

Approximately 10% of the data showed high D.O. values but low nitrate (region B of Fig. 5). These samples were generally taken from sandy areas in which there was little or no agricultural activity, though a small number of points representing very shallow samples (less than 2 m depth) from fine-textured sites also fall within this region.

Area C represents samples that showed significant concentrations of both NO_3^- and D.O. With the exception of a small number of points from shallow sampling points at fine-textured cultivated sites, the data within this zone came entirely from coarse-textured cultivated areas.

Of particular note, less than 1% of the data points fall within region D, the region of low D.O. but significant NO_3^-. Though these were all from cultivated areas, both fine- and coarse-textured geologic materials are represented.

Because of the uneven distribution of sampling points with respect to land use and geologic conditions, the absolute value of percentage occurrence in the various regions of Fig. 5 have little significance. Nevertheless, the trends are believed to be of importance. In particular, the high occurrence of points in the region of low NO_3^- and low D.O., even in sandy cultivated areas, suggests that denitrification is an active process in some ground-water environments. This is further supported by the fact that there were very few occurrences of high nitrate concentrations in regions of low D.O. This latter observation suggests that if the geochemical conditions are sufficient to remove the D.O., then once the D.O. has been depleted the conditions are sufficient for denitrification to proceed.

The trends in the data also indicate that fine-textured soils, whether cultivated or uncultivated, do not contribute to extensive nitrate contamination of ground water. While this may in part be the result of denitrification, it may simply reflect the fact that there has been insufficient time for nitrate to penetrate to significant depths in geologic materials of low permeability. From region C of Fig. 5, it is also clear that nitrate contamination of ground water in agricultural areas is common. This occurs predominantly in surficial sandy aquifers and often to concentrations that exceed the drinking water limit.

Of particular interest is the fact that sandy surficial aquifers under similar land use yield considerable data in both regions A and C of Fig. 5. This suggests that in some sandy aquifers denitrification is an active and effective sink for nitrate, while in others denitrification appears to be of little or no importance. Clearly, the controls on the occurrence of denitrification are an issue of some importance.

3 Controls on the Occurrence of Denitrification

From a further consideration of the survey data, it was recognized that the occurrence of denitrification was associated with shallow water table conditions (water table at depths of less than about two to three meters), while in situations where the water table was at a greater depth, there was generally no evidence of denitrification. Examples of these trends are shown in Fig. 6. Fig. 6a is data from a sandy site near Rodney that had been under a corn-soybean rotation for several years. The nitrate concentrations in the upper meter of the aquifer were variable over time, and as suggested in Fig. 6a, commonly reached or exceeded the drinking-water limit. Below a depth of 2 m (1 m below the water table) nitrate concentrations were consistently below detection. Dissolved oxygen showed a similar trend, declining from about 6 mg/L near the water table to nondetectable values at a depth of 1 m below the water table.

The results of the tests are shown in Fig. 7. At the Rodney site (shallow water table), the control experiment showed a decline in nitrate concentration accompanied by an increase in nitrous oxide. The calculated rate of denitrification (2.4×10^{-5} g-N/L.h) was within the range reported for the same site by Trudell et al. (1986) (7.8×10^{-6} to 1.3×10^{-4} g-N/L.h). The amended case showed a much higher rate of nitrate loss and nitrous oxide production at early time, but at times greater than about four days, the rates were similar to the control case. The period of rapid decline in nitrate concentration was shown to correspond with the period of rapid decline in dissolved organic carbon, presumably the period during which the added glucose was consumed.

The behavior at the Alliston site (deep water table) was substantially different (Fig. 7b). In the control test, there was a slight decline in NO_3^- concentration over a period of fourteen

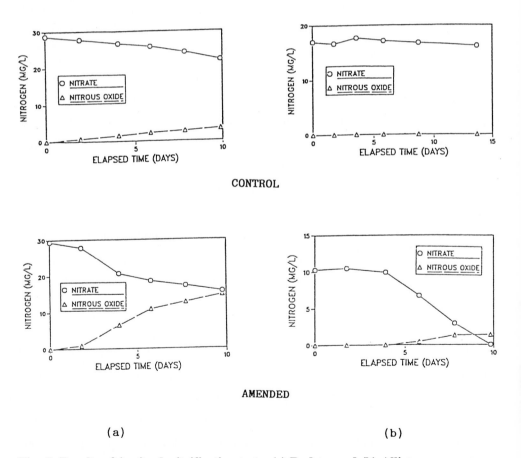

Fig. 7. Results of _in situ_ denitrification tests: (a) Rodney and (b) Alliston

days; however, this was similar to the observed decline in the nonreactive (Br⁻) tracer. There was also no detectable accumulation of N_2O. Thus, consistent with the conclusions based on the geochemical profiles, under natural conditions denitrification at this location does not appear to be an active process. With the addition of glucose (Fig. 7b "amended"), there was a lag period of about four days, followed by a rapid decline in the nitrate concentration and an accumulation of N_2O. The results provide strong evidence that the occurrence of denitrification at the deep water table site was indeed limited by the lack of labile organic carbon. Though the initial lag period may represent the time for development of the denitrifying population of bacteria, clearly the necessary bacteria were present at this depth in the aquifer.

The *in situ* tests were supplemented with laboratory microcosm tests of the soil's ability to support denitrification. These tests were conducted on core samples collected from ground surface to below the water table at both the shallow and deep water table sites. The tests showed that at both sites, there was sufficient labile organic carbon present to support significant rates of denitrification to depths of about 2 to 2.5 m below ground surface. This finding suggests that, under the conditions of these two sites, labile organic carbon persists to depths of between 2 and 3 m below ground surface. If the water table occurs at shallower depths, denitrification within the aquifer could be expected; if the water table is at greater depth, denitrification would not be expected.

The *in situ* denitrification tests and the profiles of potential for denitrification gave consistent results and, perhaps of greater importance, the results are consistent with the results of the NO_3^--D.O. survey. This provides some degree of confidence in the generality of the results obtained in the detailed tests at Rodney and Alliston.

The fact that denitrification appears to be related to leaching of dissolved organic carbon from the soil surface could have important consequences. For example, by altering management practices, it may be possible to enhance leaching and transport of organic carbon to the water table. This in turn could promote denitrification with the potential to reduce the extent and severity of nitrate contamination of the aquifer. Furthermore, in many areas, land-use restrictions are being developed as a means of protecting ground-water quality. In developing these restrictions, coarse-textured soils overlying an aquifer are considered to provide the greatest potential for nitrate contamination, and thus the use of fertilizers on these soils is being restricted. The results of the Ontario studies suggest that the depth to the water table should also be considered in the delineation of nitrate-sensitive areas. That is, if the water table is shallow, denitrification may provide a sufficient sink for nitrate such that land-use restrictions are not necessary.

4 Concluding Comments

Though there has been no comprehensive survey to determine the extent of nitrate contamination of Ontario ground waters, there is no reason to believe the conditions to be substantially different from those reported for the United States (Hallberg, 1989, for example) and western Europe (Juergens-Gschwind, 1989, for example).

In particular, aquifers that are connected to the surface by geologic materials of high permeability and that underlie areas of intensive agricultural production can be expected to have elevated nitrate concentrations. The available data suggest that in these environments, contamination can be extensive with concentrations frequently exceeding the drinking water limit of 10 mg/L NO_3-N. The lack of data is such that it is not possible to evaluate changes in the extent or severity of contamination over time.

Though a small amount of data is available, it appears that in fine-textured soils underlain by clayey geologic materials, significant nitrate concentrations do not occur at depths greater than one to two meters below ground surface. This suggests that fine-textured soils are not a significant source of nitrate contamination in ground water and that confined aquifers are unlikely to be contaminated by agricultural sources of nitrate. Though not the subject of the studies reported in this paper, it is recognized that fine-textured soils can contribute nitrate to surface waters both through surface runoff and discharge from tile drains.

Nitrate concentrations in Ontario aquifers are highly variable with depth, and in some situations denitrification appears to be a significant contributor to this variation. In particular, in hydrogeologic settings with shallow water tables (2 to 3 m below ground surface), high nitrate concentrations are generally limited to depths of 1 to 2 m below the water table. In these settings, the quality of domestic water supplies can often be improved by installing the well screen at a greater depth in the aquifer. In aquifers with deeper water tables, nitrate contamination can penetrate to considerable depths with no indication that denitrification acts as a significant sink for nitrate.

The apparent relationship between the transport of dissolved organic carbon from the soil zone to the aquifer and the occurrence of denitrification at shallow depths within the aquifer has several practical implications. In particular, enhanced transport of organic carbon could promote denitrification, thus reducing the severity of nitrate contamination. It also appears that the depth to the water table, or more likely the residence time of infiltrating water within the unsaturated zone, is a factor that should be considered in evaluating the sensitivity of land areas to nitrate contamination. This could be particularly important as various jurisdictions move towards land-use restrictions.

Recognizing the potential importance of denitrification in aquifers, it is important that much more be learned concerning the transport of organic carbon from the soil zone and through the unsaturated zone. Source, persistence, leaching rate, lability, and temporal variability in these properties are all issues that need to be addressed.

As a final comment, it is important that monitoring methods be selected that are appropriate for the objectives of a particular study. If the purpose is to determine the nitrate concentration in drinking water that a particular household is exposed to, then samples should be collected from the household supply. On the other hand, if objectives are related to the extent of aquifer contamination or the processes of nitrate transport within an aquifer, then data obtained from existing domestic wells can be very misleading. Because of the high degree of variability that appears to be characteristic of nitrate concentrations, a three-dimensional array of sampling points using multilevel sampling installations is strongly recommended.

References

Egboka ECB (1978) Field investigations of denitrification in groundwater. M.Sc. Thesis, Department of Earth Sciences, University of Waterloo, Waterloo, Ontario

Gillham RW, Starr RC, Akindunni FF, O'Hannesin SF (1984) Studies of nitrate distribution and nitrogen transformation in a shallow sandy aquifer near Alliston, Ontario. Proceedings, Technology Transfer Conference No. 5, Policy and Planning Branch, Ontario Ministry of the Environment, November 27-28, p 25

Table 2. The employed numerical values of the constants and coefficients for simulation of nutrient transport within the canals of Konya Plain

K_1	0.3 day^{-1}
K_2	Thackson and Krenkel equation for canals with normal slope; O'Connor and Dobbin's Equations for canals with high slope and canals following chutes.
K_3	0.05 day^{-1} for canals with normal slope; 0.0 for others.
Q	Manning's Equation with n = 0.02
K	Dispersion constant = 300
K_4	0.01 for the canals with low slope and canals before pumping stations; 0.0 for the others (Sediment oxygen demand)
β_1	0.55 day^{-1}
β_2	1.10 day^{-1}
β_3	0.21 day^{-1}
β_4	0.36 day^{-1}
σ_4	0.05 day^{-1}
σ_5	0.50 day^{-1}

Notes: Sewage characteristics through SİMAŞ Infrastructure Company

2. Ground-water infiltration is negligible compared to the main river flow. This assumption is reasonable since the Seyhan River travels only about 60 km downstream from Adana before it reaches the Mediterranean Sea. Further, only about 5% of the catchment area is affected. The effect of agricultural and storm runoffs was also excluded in the simulation.

3. The empirical relationships between velocity of flow, depth of flow and discharge which were obtained from General Directorate of Survey for Electrical Works Station No. 1807 are applicable along the Seyhan River downstream of Adana.

2.2.2 Assumptions Related to Konya Plain Closed Catchment

The canal is earthen and has a well-defined geometry. Detailed and accurate flow records were available from the State Water Works of Turkey, Fourth Division. Therefore the velocities and depths of flow could be accurately estimated using hydrological data and knowledge of the geometrical properties of canals.

3 Results of the Pollutant Transport Study

3.1 Seyhan River Catchment

Study results for domestic pollutants originating from Adana City are summarized in Table 3. These results reflect average hydrological and meteorological conditions for the specified seasons. The BOD_5 load to the Mediterranean Sea ranged between 500 and 1000 tons/month depending upon seasonal and hydrological conditions. The total nitrogen contribution remained about 160 tons/month, and the organic and dissolved-P load was estimated to be 28 tons/month.

Table 3. Estimated average pollutant loads to the Mediterranean Sea from the City of Adana as of 1989 (tons/month)

Parameter	Season			
	Summer	Autumn	Winter	Spring
BOD_5	488	992	1123	1221
Organic-N	33	49	52	50
NH_3-N	26	68	86	74
NO_2-N	18	23	14	21
NO_3-N	81	23	9	16
Organic-P	7	11	14	12
Dissolved-P	22	17	14	16

3.2 Konya Closed Catchment

The BOD_5, DO, Organic-N, NH_3-N, NO_2-N, NO_3-N, organic-P and dissolved-P loads to Tuz Gölü Lake were estimated using the mathematical approach described in the previous section. Table 4 summarizes the results of simulations for different seasonal hydrological conditions.

BOD loads to Tuz Gölü Lake have been estimated to range from 400 to 700 tons/month. These estimates are based upon hydrological and climatological conditions in 1988, a period during which no water pollution control practices were in use. The measured BOD flux to Tuz Gölü ranged between 300 and 1100 tons/month as reported by an independent governmental organization, the State Water Works of Turkey.

Table 4. Estimated present pollutant loads to Tuz Gölü as of 1988 (tons/month)

Parameter	Winter	Flood	Low Flood
BOD	707	702	445
Organic-N	39.19	37.82	28.24
NH_3-N	70.73	62.57	54.02
NO_2-N	11.57	15.20	19.03
NO_3-N	10.07	16.62	28.54
Organic-P	10.47	9.55	5.93
Dissolved-P	13.06	13.08	16.78

The estimated (NH_3-N + NO_2-N + NO_3-N) flux was about 100 tons/month irrespective of the hydrological conditions and the season. The measured fluxes ranged between 55 and 247 tons/month for the last five years. The high values indicate the possibility of nitrogen sources from other than domestic origin.

Estimated (organic-P + dissolved-P) loads to Tuz Gölü ranged between 14.6 and 17.7 tons per month depending upon hydrological and seasonal conditions. Unfortunately, no measurements of Organic-P and Dissolved-P were available for comparison with model predictions.

4 Conclusions

The equations predicted the BOD_5 loads to Tuz Gölü Lake fairly closely. The orders of magnitude of predictions for (NH_3-N + NO_2-N + NO_3-N) flux were the same as the actual measurements. Some high loads of (NH_3-N + NO_2-N + NO_3-N) which were observed were attributed to nondomestic sources.

References

Brown L, Barnwell Jr TO (1987) The enhanced stream water quality models. USA-EPA
Master plan for Konya wastewater collection system (1987) (In Turkish), Simaş, Ankara Turkey
Mean monthly flow data for 1935-1980 (1983) (In Turkish) General Directorate of Survey for Electrical Works, Turkey
O'Connor DJ, Dobbins WE (1958) Mechanisms of reaeration, ASCE J of San Eng Div 123:641-684

Personal communications with the authorities of İller Bankası of Turkey (1989)

Soyupak S, Sürücü G (1988) An evaluation of treatment alternatives for Konya City wastewaters (In Turkish) METU, Ankara, Turkey

Soyupak S (1989) Development of approaches for the continuous assessment of pollutant loads, WHO project No:ICP/CEH 047 Tur(14 B)

Soyupak S, Sürücü G (1989) Pollutant transport for Tuz Gölü, Int J of Env Studies 33:291-298

Thackson EL, Krenkel PA (1969) Reaeration prediction in natural streams, ASCE J of San Eng Div 95:65-94

Turkey, Ministry of Energy and Natural Resources (1988) Plain of Konya, main drainage channel, pumping stations, monthly records (In Turkish), DSİ, 4th Division Directory, Konya, Turkey

Turkey, Ministry of Energy and Natural Resources (1988) Water quality records (In Turkish) DSİ, Ankara Turkey

Turkey, Ministry of Finance and Customs, (1988) Personal communications with the authorities

Turkey, State Institute of Statistics (1987) Census of population (In Turkish) DİE

POTENTIAL NITRATE POLLUTION OF GROUND WATER IN LIMESTONE TERRAIN BY POULTRY LITTER, OZARK REGION, U.S.A.

K.F. Steele and W.K. McCalister
Arkansas Water Resources Research Center
and Department of Geology
University of Arkansas
Fayetteville, Arkansas 72701 U.S.A.

Abstract

The Ozark Region of Arkansas is the major poultry-producing area of the United States. Large quantities of poultry waste are spread as fertilizer on thin soils of pastureland overlying limestone aquifers. Because these aquifers provide domestic water supplies for the rural population and are susceptible to contamination from surface water, there is concern that nitrate leached from poultry litter may be polluting the ground water. In response to this concern, well water from a major poultry-producing area was compared with that from a forested area. Although the nitrate concentration of the well water from the poultry-producing area (2.83 mg/L as nitrogen) is considerably below the drinking-water limits of 10 mg/L set by the U.S. Environmental Protection Agency (U.S. EPA), it is about ten times that of springs in the forested area. Variations in nitrate concentrations in the ground water are dependent upon the availability of nitrogen and recharge.

1 Introduction

Although nitrate contamination of water by commercial fertilizers and feed lots has been extensively investigated (e.g., Beck et al., 1985; McLeod and Hegg, 1984; Hill, 1982; Khaleel et al., 1980; Spalding et al., 1978; Sommerfeldt et al., 1973; Lorimar et al., 1972; Gillham and Webber, 1969), very little research has been conducted on the effects of land application of poultry litter (Adamski and Steele, 1988; Wolf et al., 1988; Giddens and Barnett, 1980; Liebhardt et al., 1979). Arkansas is the national leader in broiler production and in 1988 produced over 900 million broilers, turkeys and hens (Arkansas Agricultural Statistics Service, 1989).

A majority of Arkansas' poultry production is in the Ozark Region of the northwestern portion of the state. Fractures in the limestone combined with the thin soils of this region

NATO ASI Series, Vol. G 30
Nitrate Contamination
Edited by I. Bogárdi and R. D. Kuzelka
© Springer-Verlag Berlin Heidelberg 1991

make the shallow limestone aquifers susceptible to contamination from surface sources (Fig. 1). Surface application of poultry litter (manure and associated bedding material such as sawdust) to pastureland as fertilizer has raised concern that nitrate from the litter and manure from grazing cattle could pollute the ground water.

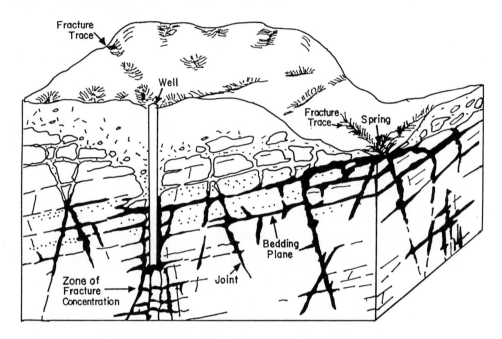

Fig 1. Schematic diagram typical of the Boone-St. Joe aquifer showing the relationship of a spring and a high-yield water well to fracture and bedding planes

Two studies specifically designed to analyze the effects of land-applied poultry wastes on ground-water quality in northwestern Arkansas have been conducted. One study focused on springs issuing from the Boone-St. Joe Formation during 1986-1987 (Adamski and Steele, 1988), and the second study in 1989 (reported here) focused on wells completed in the Boone-St. Joe aquifer.

2 Study Area and Land Use

The northeastern Washington County area was used for the present study and for the earlier study by Adamski and Steele (1988) (Fig. 2) to investigate the effect of poultry litter on nitrate concentrations of ground water because: (1) it is one of the highest-density poultry-producing areas in the United States and (2) the shallow Boone-St. Joe limestone aquifer underlying the area is susceptible to contamination from surface sources.

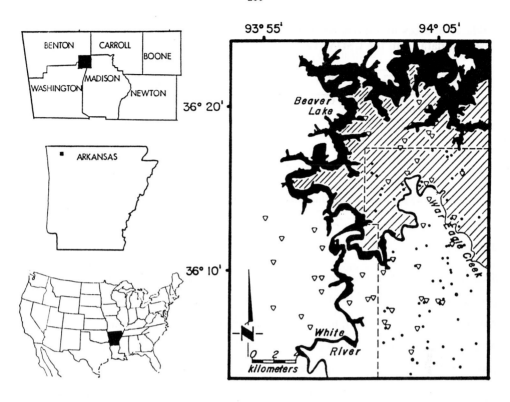

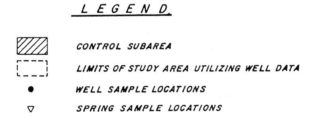

Fig. 2. Location of well sites for this study and for springs used by Adamski and Steel (1988)

For both studies, experimental and control subareas were defined (Fig. 2). The two subareas are adjacent and, with the exception of land-use practices, have similar geology, meteorology and hydrogeology. In 1989 there was over 44×10^6 kg of wastes (0.5×10^6 kg of nitrogen) (Table 1) produced in the 140 km^2 experimental subarea (Fig. 2). The majority of the wastes and nitrogen was from poultry litter. A 145 km^2 portion of an adjacent forested wildlife management area was used as the control subarea (shaded portion of Fig. 2).

The smaller difference between control and experimental wells compared to that for springs is probably the result of some contamination of the control wells (Table 2). The wells used for this study were domestic wells which may be contaminated by runoff from barnyards, lawns and/or by septic tank effluent. This observation suggests: (1) that it is difficult to obtain "true" control wells in a relatively shallow limestone aquifer because of anthropogenic effects, and (2) that some of the ground-water contamination in the experimental subarea is from sources other than poultry litter and cattle manure. Data on nitrate concentrations of spring water (0.14 to 0.33 mg/L) from other relatively pristine areas in the area (mostly forested with very low population) with similar hydrogeology (Table 3), confirms that samples of ground water from the control subarea should be less than 0.40 mg/L nitrate rather than the 1.62 and 1.78 mg/L observed for control wells (Table 2).

Table 3. Comparison of seasonal nitrate concentrations of springs from the control subarea with springs from other pristine areas. Concentations are mg/L nitrate+nitrite expressed as nitrogen. The numbers in ( ) are the number of sites collected. Control subarea data are for the three seasons in Table 1

	Other Pristine Areas of the Region		
Control Subarea	Ponca[1]	Rush[1]	Zinc[1]
0.40 (18)	0.14	0.33	0.30
0.16 (10)	(48)	(52)	(43)
0.02 (8)			

[1] Data for these areas from Steele (1983).

Although there are statistically higher concentrations of nitrate in ground water from experimental subareas (about ten times that of control springs based on weighted mean averages), these concentrations are considerably below the drinking water limit of 10 mg/L set by the U.S. EPA (1985). There is concern that the soil in most of northwestern Arkansas has more than sufficient available nitrogen present for vegetation growth, and therefore it is probable that much of the nitrogen in any additional litter applied to the land may be leached into the ground water and significantly increase nitrate concentrations.

Another report (Steele et al., 1990) compares the nitrate concentrations of the shallow (Boone-St. Joe) and deeper (Everton) limestone aquifers. Nitrate concentrations for the deeper

Everton aquifer (1.5 mg/L) are about half those for the shallow Boone-St. Joe aquifer (2.8 mg/L) and are statistically significant (0.05 alpha, Wilcoxon two-sample test). The difference between the two aquifers is attributed to the separation of the two aquifers by about 18 meters of relatively impermeable Chattanooga Shale Formation.

7 Comparison of Nitrate Concentrations from Wells and Springs

Table 2 indicates that springs had slightly higher nitrate concentrations than wells in the experimental subarea during the spring and fall seasons, and wells had higher concentrations during the winter. These differences are probably not meaningful because spring and well samples were collected in different years (1986-1987 and 1989, respectively) and environmental conditions (for example, timing and amount of litter application and amount of recharge) could have been different for the two study periods. As noted previously, for the shallow Boone-St. Joe aquifer the proximity of control wells to human activities may result in more contamination of wells than springs because control springs were not located near human activity.

8 Nitrate Concentration Variations

Spring is the season expected to have the highest ground-water nitrate concentrations because this is the season when: (1) most of the poultry litter is applied to the land, (2) heavy spring rains cause major recharge to the aquifer and (3) there is little nitrate uptake by vegetation (which is still mostly dormant). It is interesting to note that the spring season indeed has the highest average nitrate concentration for both wells and springs (Table 2), even though samples were collected during different years. Despite this logical explanation for higher nitrate values occurring in the spring, comparison of the well data by the nonparametric Kuskal-Wallis statistical test (0.05 alpha) supports the null hypothesis that there are no significant differences among the seasons. The Kuskal-Wallis test was used rather than the Wilcoxon two-sample test because it is a multiple-sample test that allows simultaneous comparison of all three seasons.

Two springs in the experimental subarea exhibited decreases in nitrate concentration (3 to 2 mg/L and 6 to 1 mg/L) following initiation of rain during an October 1987 storm. During the same period, a control spring maintained nitrate values less than 0.2 mg/L (Steele et al., spring water chemistry in a limestone terrane. In: Proceedings of the Southern Regional Ground Water Conference. Nat Water Well Assoc, Worthington Ohio, 50-66

Steele KF, McCalister WK, Adamski, JC (1990) Nitrate and bacteria contamination of limestone aquifers in poultry/cattle producing areas of northwestern Arkansas, USA. Environmental Contamination 238-241

US Department of Agriculture (1975) Agricultural Waste Management Field Manual. US Gov Printing Office, Washington DC Chapters 4 and 11

US Environmental Protection Agency (1983) Methods for Chemical Analysis of Water and Wastes. Environmental Protection Agency, Environmental Monitoring and Support Laboratory, Cincinnati Ohio, 430

US Environmental Protection Agency (1985) National primary drinking water regulations: synthetic organic chemicals, inorganic chemicals and microorganisms; proposed rule. Fed Register 50:46934-47022

Van den Huevel, P. (1979) Petrography of the Boone Formation, Northwest Arkansas. Unpublished MS Thesis, University of Arkansas, Fayetteville, Arkansas

Wolf DC, Gilmour JT, Gale PM (1988) Estimating potential ground and surface water pollution from land application of poultry litter-II. Pub No 137, Ark Water Resources Research Center, University of Arkansas, Fayetteville, Arkansas

NITROGEN MANAGEMENT RESEARCH IN THE PRESIDENT'S WATER QUALITY INITIATIVE

C.A. Onstad, J.S. Schepers[1] and W.E. Larson[2]
Deputy Area Director for the Midwest Initiative on Water Quality
USDA-Agricultural Research Service
North Iowa Avenue
Morris, Minnesota 56267 U.S.A.

Abstract

Contamination of our ground and surface-water supplies from normal use of pesticides and fertilizers is a growing national concern. Efforts from both the public and private sectors are required to develop safer pest controls and to adopt environmentally sensitive farm production practices. The President's Water Quality Initiative is based on three policy objectives (Bush, 1989): First, the federal government is committed to protecting ground-water resources from fertilizer and pesticide contamination without jeopardizing the economic vitality of U.S. agriculture. Second, water-quality programs must be designed to accommodate both the immediate need to halt contamination and the future need to alter fundamental farm production practices that result in contamination. Third, farmers must ultimately be responsible for changing production practices to avoid contaminating ground and surface waters. Federal and state resources can provide valuable information and technical assistance to producers so that environmentally sensitive techniques can be implemented at minimum cost.

1 Introduction

Agriculture's challenges in the initiative are first, to conduct the bio-chemical-physical research needed to understand (and improve the management of) the fate and transport of chemicals used in agriculture; second, to develop environmentally sensitive alternative cropping systems and components; third, to educate, demonstrate, and deliver alternative practices to farmers and ranchers; and fourth, to monitor implementation of alternative systems and components on the land.

---

[1] Soil Scientist, USDA-Agricultural Research Service, Agronomy Department, University of Nebraska, 113 Keim Hall, Lincoln, NE 68583-0915, U.S.A.
[2] Professor Emeritus, Soil Science Department, University of Minnesota, 439 Borlaug Hall, St. Paul, MN 55108 U.S.A.

NATO ASI Series, Vol. G 30
Nitrate Contamination
Edited by I. Bogárdi and R. D. Kuzelka
© Springer-Verlag Berlin Heidelberg 1991

extensively for many high-value crops where the time required and added costs can be justified to grow a product that is appealing to the public. Tissue testing in corn production systems has been used primarily to diagnose secondary element and micronutrient deficiencies. Over time, critical and toxic levels have been established for most crop nutrients, including N. Adoption of this technology to improve N management practices for corn production has been slow because critical levels change with growth stage, and remedies to correct an N deficiency may be impractical. The practical approach to crop N nutrition used by many producers is to make sure that a deficiency does not develop. Unfortunately, this philosophy has lead to overfertilization and contributed to the current situation where large areas of corn producing states now have ground-water nitrate-N concentrations above the recommended drinking water standard of 10 mg/L.

3 Management Effects

Cause and effect relationships between agricultural practices and occurrence of NO_3-N in ground water are difficult to document. Reasons for this difficulty are that it may take several years or more for NO_3-N to leach to the aquifer, sources of NO_3-N are difficult to identify, and soil conditions conducive to leaching are usually not well-documented. Nevertheless, correlations between NO_3-N concentration in ground water and water-quality related parameters (i.e., depth to aquifer, irrigation well density, soil type, livestock density, etc.) help identify various production practices that should be targeted for more critical management (Muir et al., 1973). One such example is from the Platte River Valley in central Nebraska (Schepers et al., 1991) representing 200,000 ha of irrigated corn where increased levels of NO_3-N in irrigation well water are positively correlated with applications of N fertilizer above the recommended rate (Fig. 1). The apparent overapplication of N fertilizer, which probably has been occurring for at least three decades, has resulted in elevated levels of ground-water NO_3-N. The strong relationship between NO_3-N concentrations in irrigation water and the apparent overapplication of N fertilizer implies that crop N availability was probably excessive during at least portions of the year and that subsequent leaching occurred.

Intuitively, concern for NO_3-N leaching is probably greatest during the growing season, especially under irrigated conditions. It should be equally obvious that surplus NO_3-N in soil is also subject to leaching during the winter fallow period and may present a greater threat

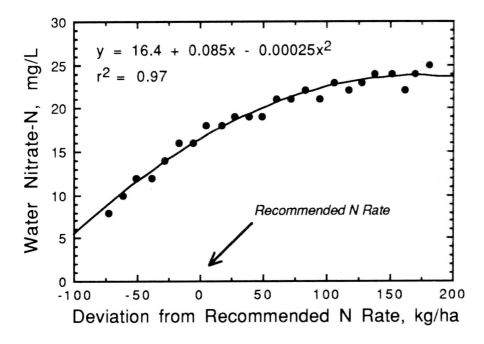

$$y = 16.4 + 0.085x - 0.00025x^2$$
$$r^2 = 0.97$$

Fig. 1. Effect of fertilizer N application rate within the Central Platte Natural Resource District's ground-water management area on NO_3-N concentration in irrigation water

to ground-water quality than leaching during the growing season, depending on climatic conditions. For example, high levels of residual N in some Platte River Valley fields before the 1988 cropping season undoubtedly reflected excess fertilizer N applied in 1987. It is probably reasonable to assume that production practices were similar during 1987 and 1988. As such, many of the same fields that received excess N fertilizer in 1988 would have also been overfertilized in 1987. It then follows that producers who applied excess N fertilizer might be expected to have the highest levels of N carryover into the next growing season (Fig. 2) and also have the highest levels of NO_3-N in ground water (Fig. 1). Therefore, it seems logical that efforts to reduce NO_3-N leaching must begin by addressing fertilizer strategies and water-management practices that ultimately improve N-use efficiency.

Fertilizer N management strategies used by producers are influenced by basic considerations such as soil type, cropping system, and climate. Other considerations include equipment availability, cost of various fertilizer N sources, labor requirements, and the degree of risk associated with developing an N deficiency. For crops such as corn, the basic risk consideration is that of yield reduction associated with an N deficiency. It is generally recognized that yield reductions attributed to an N deficiency are unacceptable because of current prices of corn grain and N fertilizer. Producers can break even by applying an extra 0.45 kg/ha N fertilizer if corn yields increase by 1 kg/ha. At the break-even point, a 1 kg/ha increase in yield removes approximately an extra 15 g N/ha in grain and 10 g N/ha in stover. Combining the additional N uptake in grain and stover results in a net fertilizer-use efficiency of approximately 2.5% at the break-even point. The environmental implication is that much of the N not taken up by the crop is subject to leaching and eventual ground-water contamination. Therefore, producers who strive to break even with the last unit of additional fertilizer N application are subjecting the environment to a large risk.

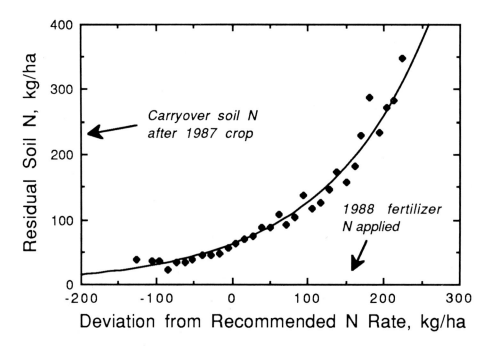

Fig. 2. Relationship between fertilizer N application rate within the Central Platte Natural Resource District's ground-water management area and residual soil N

Risk assessment in agricultural production is uncertain because it involves a number of unpredictable variables. Foremost of these uncertainties is climate. Excess precipitation enhances leaching and can result in crop N deficiency and reduced yields. To guard against this possibility, producers frequently apply some additional N fertilizer as insurance against yield reduction. The concept of insurance N is only part of a producer's N management strategy. Tradition and past production experiences have a strong influence on the amount of fertilizer N applied by producers. This is illustrated by 1988 data from the Central Platte River Valley of Nebraska, where producers tended to apply 150 to 200 kg N/ha regardless of the amount of N recommended (Fig. 3).

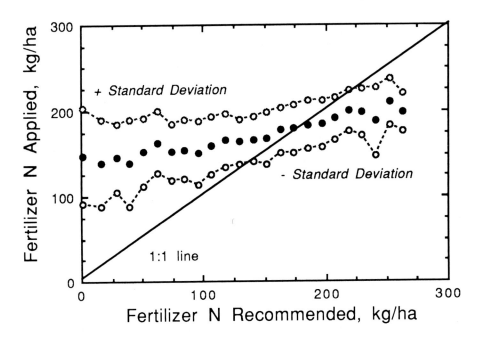

Fig. 3. Fertilizer N application rates as related to recommended amounts for irrigated corn production within the CPRND Phase II Ground Water Protection Area for 1988

The University of Nebraska's fertilizer N recommendations were based on soil test data. However, nearly stable fertilizer N application rates across a wide range of fertilizer N recommendations suggests that producers generally disregard soil test information. An alternate interpretation of the data is that although producers recognize the value of soil testing, concern for yield reductions because of possible N losses leads them to apply fertilizer N at levels consistent with past experience. Other factors also influence producer decisions. For example, it is doubtful that the entire amount of residual soil N in the root zone at the beginning of the growing season will be equally available to the crop. As producers attempt to compensate for inefficiencies and losses for various sources of N, they must evaluate difficult tradeoffs between economic and environmental considerations.

4 Conclusion

The role of legislation and subsequent regulations versus accelerated educational efforts and voluntary improvements in management practices to minimize nitrate leaching is uncertain. The goal of the President's Water Quality Initiative is to evaluate packages of improved management practices that have potential to reduce nitrate and pesticide leaching in an effort to protect ground-water resources. Management Systems Evaluation Area (MSEA) projects will play an important role in determining the direction of future water-quality policy in the United States.

References

Bush G (1989) President of the United States of America. Building a better America. Supplement to the State of the Union message, enhancing water quality, section g, Government Printing Office, p 92

Muir J, Seim EC, Olson RA (1973) A study of factors influencing the nitrogen and phosphorus contents of Nebraska waters. J Environ Qual 2:466-470

Schepers JS, Moravek MG, Alberts EE, Frank KD (1991) Cumulative effects of fertilizer and water management on nitrate leaching and ground water quality. J Environ Qual (in press)

U.S. EPA (1990) National Pesticide Survey: Phase I Report. Executive Summary. November.

PANEL DISCUSSION: NITRATE EXPOSURE IN AGRICULTURE

D.G. Watts, Panel Chair and Reporter
Biological Systems Engineering
230 Chase Hall
University of Nebraska-Lincoln
Lincoln, Nebraska 68583-0726 U.S.A.

Panel Members: R. Mull,[1] H.P. Nachtnebel,[2] L.S. Pereira,[3] P. Rijtema[4]

1 Opening Comments by WATTS

Agriculture is one of the primary sources of nonpoint nitrate contamination of ground water. We now must deal with the fundamental issue of how to manage agriculture so as to maintain profitability for the farmer while protecting ground-water supplies from contamination for the benefit of everyone. In some situations, adjustment of agricultural management practices may be sufficient to significantly reduce contaminant emissions from agricultural lands. In other cases, management changes will not be sufficient; major conflict between farmers and other ground-water users seems inevitable. There are some locations where it may not be possible to maintain a profitable agriculture and adequately protect ground-water resources.

Each of the panelists was asked to respond to three questions:

1. Once agricultural exposure controls are implemented, are the response times so long as to make the controls unrealistic?

2. Is the best management practice a compromise between nitrate risk reduction and agricultural revenue?

3. Can the effect of agricultural practice on nitrate exposure be predicted with reasonable accuracy?

---

[1] Institute of Water Resources, Hydrology and Agricultural Engineering, University of Hannover, Appelstraße 9A, 3000 Hannover 1, Germany.
[2] Institute of Water Resources Management, Universität für Bodenkultur, Gregor Mendel-Str. 33, A-1180 Wien, Austria.
[3] Department of Agricultural Engineering, Technical University of Lisbon, Faculty of Agriculture, Tapada da Ajuda, 1399 Lisboa Codex, Portugal.
[4] The Winand Staring Centre for Integrated Land, Soil and Water Research, P.O. Box 125 6700 AC Wageningen, The Netherlands.

NATO ASI Series, Vol. G 30
Nitrate Contamination
Edited by I. Bogárdi and R. D. Kuzelka
© Springer-Verlag Berlin Heidelberg 1991

III. HEALTH CONSEQUENCES

EVIDENCE THAT N-NITROSO COMPOUNDS CONTRIBUTE TO THE CAUSATION OF CERTAIN HUMAN CANCERS

M. Crespi[1] and V. Ramazzotti
Department of Environmental Carcinogenesis, Epidemiology and Prevention
National Cancer Institute "Regina Elena"
viale Regina Elena, 291
00161 Rome, Italy

Abstract

N-nitroso compounds (NOC) are potent animal carcinogens. Their effect on humans is still inconclusive. *In vivo* formation is one of the most likely means of exposure and possibly is linked with gastric carcinogenesis. Research with the NPRO test was performed by endoscopy and urine and gastric juice collection to establish the nitrosation potential in normal subjects and in different pathological conditions. Other cancer sites possibly correlate with environmental and occupational exposure to NOC. Guidelines for prevention are proposed.

1 Introduction

Laboratory studies involving some 40 animal species and focusing on target organs including the stomach, esophagus, nasal cavity, lung, liver, and kidney have established that the majority of N-nitroso compounds (NOC) are carcinogens in experimental animals (Magee and Barnes, 1956; Sugimura and Fujimura, 1967; Magee et al., 1976; Adamson and Sieber, 1980; Bogovski and Bogovski, 1981; Archer and Labuc, 1982; Hodgson et al., 1982; Labuc and Archer, 1982; Preussmann, 1977; Schmahl and Scherf, 1983). Direct evidence of a carcinogenic effect of these chemicals is not yet available for humans, but many observations suggest that NOC play an important role as possible "initiators" in human carcinogenesis.

The mechanism postulated for the carcinogenic effect consists of the formation *in vivo*, by metabolic activation, of reactive alkylated intermediates with DNA. Their carcinogenicity and cytotoxicity, though, may also develop by other mechanisms which do not involve DNA

---

[1] The original experimental work reported in this paper was performed in collaboration with H. Ohshima, N. Munoz and H. Bartsch of the International Agency for Research on Cancer, Lyon, France and V. Casale and A. Grassi of the National Cancer Institute "Regina Elena," Rome, Italy under a research contract of the Italian National Research Council (CNR).

alkylation (Magee, 1987). N-nitroso compounds interact with the human organism through either exogenous or endogenous exposure. Occupational or life-style factors account for most exogenous exposure (Preussmann, 1984). Endogenous exposure, resulting from an *in vivo* formation, probably represents the greater source of exposure for the general population even though most of the responsible compounds have not yet been chemically characterized (Mirvish, 1975a; Michejda et al., 1982).

Possible relationships between endogenously formed NOC and human cancer have been postulated for the following reasons:

1. Exposure to nitrosatable amines and to nitrosating agents is widespread and continuous in the human environment.

2. *In vivo* formation (intragastric in particular) of NOC has been demonstrated, but many intrinsic and dietary factors influence the process. This may explain some discrepancies in the epidemiological data dealing with exposure to NOC and their precursors and cancer frequency (Sander and Seif, 1969; Mirvish et al., 1973; Ruddel et al., 1976; Forman et al., 1985; Challis et al., 1984).

2 Exposure to NOC and Cancer of the Digestive System

More evidence has been assembled for gastric cancer than for other potential cancer sites establishing a relationship between NOC exposure and cancer incidence. Among the epidemiological evidence is the presence of high levels of nitrite and/or nitrate in the soil, water and food in high-incidence areas of gastric cancer (Armijo and Coulson, 1975; Amadori et al., 1980; Armijo et al., 1981; National Research Council, 1981). Clinical evidence includes the presence of precancerous conditions which promote the *in vivo* formation of NOC (Blackburn et al., 1968; Stalsburg and Taksdal, 1971; Cuello et al., 1976; Schlag et al., 1980; Reed et al., 1981). Among these conditions are, for example, chronic atrophic gastritis (CAG), pernicious anemia and gastric resection. Such conditions, as well as treatment with drugs which block acid secretion, cause an increase in the gastric pH, thus creating a favorable environment for the development of nitrate-reducing bacteria in the stomach. These bacteria are able to transform nitrate to nitrite, with a consequent increase of endogenous nitrosation (Correa et al., 1975; Ruddel et al., 1976; Schlag et al., 1980; Reed et al., 1981; Charnley et al., 1982; Elder et al., 1982; Reed et al., 1982; Mirvish, 1983).

Consider, for example, the biological steps involved in the *in vivo* formation of nitrosamines from nitrite: nitrites (NO_2^-), preformed or consequent to the reduction of nitrates (NO_3^-), by reaction with amines (secondary or tertiary) form nitrosamines (Druckrey and Preussmann, 1962; Mirvish et al., 1975a; Kunisaki and Hayashi, 1979). This process takes place mostly in the stomach, especially when certain pathological conditions are present, but also occurs in the oral cavity and probably in the large intestine (Hartman, 1982; Suzuki and Mitsuoka, 1984).

Intragastric nitrosation is influenced by the following factors:
- The concentration of preformed NO_2^- reaching the stomach.
- The level of bacterial contamination of the stomach (total and NO_3^- reducing bacteria).
- The concentration of amine precursors.
- The presence of predisposing conditions such as chronic atrophic gastritis (CAG), or gastric resection, etc.
- The pH of the gastric juice.
- The rate of stomach filling and emptying.
- The rate of absorption and the chemical reactivity of the NOC.
- The presence of catalysts or inhibitors of the process. Catalysts include bacterial enzymes, bromides, and thiocyanates (present in the saliva, especially of smokers); inhibiting agents include polyphenols, vitamin E, and, in particular, vitamin C (Mirvish, 1975b; Ohshima et al., 1982a; Hoffmann and Brunnemann, 1983; Bartsch and Montesano, 1984; Archer, 1984; Ladd et al., 1984; Reed et al., 1984; Suzuki and Mitsuoka, 1984; Wagner et al., 1984).

Possible relationships between nitrosation, precancerous lesions and gastric cancer are presented in Table 1. This diagram is supported in part by epidemiological studies which show a direct relationship between the consumption of NOC, nitrate and nitrite on the one hand, and the prevalence and incidence of CAG and cancer of the stomach on the other (Correa et al., 1975). On these bases, a research program on the role of nitrosamines in gastric carcinogenesis was initiated at the Regina Elena Institute of Rome in collaboration with the International Agency for Research on Cancer (Crespi et al., 1987).

Table 3. Urinary excretion of N-nitrosamino acids (µg/24 h; median and range) and p values of comparison (from Crespi et al., unpublished data)

Study Area	No. of Subjects	Volume of 24-h urine/mL	NPRO	NMTCA	NTCA	SUM
High-risk area (RSM)						
Group AA (basal)	15	1100	1.11 (0.60-13.72)	0.59 (0-14.40)	3.24 (0-13.94)	7.01 (1.30-31.09)
Group AB (proline)	15	1150	1.13 (0.30-6.93)	1.22 (0-61.91)	3.69 (0-194.25)	8.99 (0.66-261.17)
Group AC (proline + vit. C)	15	1300	1.58 (0.67-7.28)	1.73 (0-12.85)	4.71 (0-13.20)	10.99 (0.92-28.91)
Low-risk area (RM)						
Group IA (basal)	15	1200	2.95 (0.44-28.68)	0.85 (0-21.15)	4.55 (0-16.00)	11.06 (2.98-65.34)
Group IB (proline)	15	1250	4.67 (0-12.76)	1.07 (0-36.11)	4.42 (0.2-36.75)	11.06 (2.49-58.88)
Group IC (proline + vit. C)	15	1200	2.99 (0.11-13.09)	2.09 (0-143.20)	5.07 (0-33.97)	12.69 (0.33-173.36)

	NPRO	NMTCA	NTCA	SUM
AA vs IA	$p < 0.05$	NS	NS	NS
AB vs IB	NS	NS	NS	NS
AC vs IC	NS	NS	NS	NS
AA vs AB	NS	NS	NS	NS
AA vs AC	NS	NS	NS	NS
AB vs AC	NS	NS	NS	NS
IA vs IB	NS	NS	NS	NS
IA vs IC	NS	NS	NS	NS
IB vs IC	NS	NS	NS	NS

NPRO: N-nitrosoproline; NMTCA: N-nitroso-2-methylthiazolidine-4-carboxylic acid; NTCA: N-nitrosothiazolidine 4-carboxylic acid; SUM: sum of N-nitrosamino acids

Kamiyama et al. (1987) used a similar protocol to study urinary excretion of N-nitrosaminoacids and nitrate by inhabitants of high- and low-risk areas for stomach cancer in northern Japan. After intake of proline, the NPRO level increased significantly only in subjects from the high-risk area; intake of vitamin C inhibited the increase of NPRO and lowered the levels of other nitrosamino acids only in the high-risk subjects. In contrast, urinary nitrate was higher in the low-risk group than in the high-risk group. Nitrate levels correlated well with vegetable consumption. Although nitrate intake by subjects in the high-risk area was thus lower, the potential for endogenous (possibly intragastric) nitrosation was higher, suggesting that the low-risk subjects already ingest agents that suppress endogenous nitrosation. Vegetables in fact contain not only nitrate but also vitamin C, which inhibits nitrosation (Archer, 1984).

Another example of the use of NPRO test, performed as an index of cancer risk, was the correlation study between urinary excretion of NOC and cancer mortality in 1035 healthy male subjects living in 26 counties in China (Chen et al., 1987). The results showed a moderate positive association of the nitrosating potential only with esophageal cancer. As expected, a negative association with background ascorbate levels was found for several cancer sites although this was significant only for esophageal cancer. The significant associations were explained by the inclusion of Song Xian county, which has a very high oesophageal mortality rate, in the study group.

As another example, Srianujata et al. (1987) found higher levels of nitrate and NPRO in Thai subjects of high risk for cholangiocarcinoma who were infected with a liver fluke (Opisthorchis viverrini) which sharply increases nitrosoproline yields.

3 Tobacco and Betel-Quid Carcinogenesis

Tobacco products contain a complex mixture of compounds, including tobacco-specific nitrosamines (TSNA). Among them, 4-(N-nitrosomethylamino)-1-(3-pyridyl)-1-butanone (NNK) and N-nitrosonornicotine (NNN) have been conclusively established as carcinogenic in laboratory animals. Table 4 lists several cancers which may be caused by exposure to tobacco products. The evidence from our studies shows that smokers tend to excrete higher levels of nitrosaminoacids than nonsmokers. A study of the combined effects of cigarette *vivo* and *in vitro*. Ann. N.Y. Acad Sci 258:175-180

Mirvish SS (1983) The etiology of gastric cancer. Intragastric nitrosamide formation and other theories. J Natl Cancer Inst 71:629-647

Nair J, Nair UJ, Ohshima H, Bhide SV, Bartsch H (1987) Endogenous Nitrosation in the oral cavity of chewers while chewing betel quid with or without tobacco. In: Bartsch H, O'Neill IK, Schulte-Hermann R (eds) The relevance of N-nitroso compounds to human cancer exposures e mechanisms. IARC Scientific Publications No.84,International Agency for Research on Cancer. Lyon, p 465

National Research Council (1981) The health effects of nitrate, nitrite and N-nitroso compounds. Part I of a 2-part study by the Committee on Nitrite and Alternative Curing Agents in Food, Assembly of Life Sciences. National Academy Press, Washington D.C.

Ohshima H, Bartsch H (1981) Quantitative estimation of endogenous nitrosation in humans by monitoring N-nitrosoproline excreted in the urine. Cancer Res 41:3658-3662

Ohshima H, Bereziat JC, Bartsch H (1982a) Monitoring N-nitrosamino acids excreted in the urine and faeces of rats as an index for endogenous nitrosation. Carcinogenesis. 3:115-120

Ohshima H, Pignatelli B, Bartsch H (1982b) Monitoring of excreted N-nitrosaminoacid as a new method to quantitate endogenous nitrosation in humans. In: Magee PN (ed) Nitrosamines and human cancer. Banbury Report 12, Cold Spring Harbor Laboratory, p 287

Ohshima H, O'Neill IK, Friesen M, Pignatelli B, Bartsch H (1984) Presence in human urine of new sulfur-containing N-nitrosamino acids: N-nitrosothiazolidine 4-carboxylic acid and N-nitroso 2-methylthiazolidine 4-carboxylic acid. In: O'Neill IK, Von Borstel RC, Miller CT, Long J, Bartsch H (eds) N-nitroso compounds: Occurrence, biological effects and relevance to human cancer. IARC Scientific Publications No. 57, International Agency for Research on Cancer. Lyon, p 77

Preston-Martin S, Yu MC, Benton B, Henderson BE (1982) N-nitroso compounds and childhood brain tumors: case-control study. Cancer Res. 42: 5240-5245

Preussmann R, Schmahl D, Eisenbrand G (1977) Carcinogenicity of N-nitrosopyrrolidine: dose response study in rats. Z Krebsforsch 90:161-166

Prokopczyk B, Brunnemann KD, Bertinato P, Hoffmann D (1987) The role of N-(nitrosomethylamino)propionitrile in betel-quid carcinogenesis. In: Bartsch H, O'Neill IK, Schulte-Hermann R (eds) The relevance of N-nitroso compounds to human cancer exposures and mechanisms. IARC Scientific Publications No. 84, International Agency for Research on Cancer. Lyon, p 470

Reed PI, Smith PLR, Haines K, House FR, Walters C (1981) Gastric juice N-nitrosamines in health and gastro-intestinal disease. Lancet II:550-555

Reed PI, Haines K, Smith PLR, Walters CL, House FR (1982) The effects of cimetidine on intragastric nitrosation in man. In: Magee PN (ed) Nitrosamines and human cancer. Banbury Report 12, Cold Spring Harbor Laboratory, p 351

Reed PI, Summers K, Smith PLR, Walters CL, Bartholomew BA, Hill MJ, Venit S, House FR, Hornig D, Bonjour JP (1984) Effect of gastric surgery for benign peptic ulcer and ascorbic acid therapy on concentrations of nitrite and N-nitroso compounds in gastric juice. In: O' Neill IK, Von Borstel RC, Miller CT, Long J, Bartsch H (eds) N-nitroso compounds: Occurrence, biological effects and relevance to human cancer. IARC Scientific Publications No. 57, International Agency for Research on Cancer. Lyon, p 975

Reichart P, Mohr U (1987) Correlation between chewing and smoking habits and precancerous lesions in hill tribes of northern Thailand. In: Bartsch H, O'Neill IK,

Schulte-Hermann R (eds) The relevance of N-nitroso compounds to human cancer exposures and mechanisms. IARC Scientific Publications 84, International Agency for Research on Cancer. Lyon, p 463

Ruddel WSJ, Bone ES, Hill MJ, Blendis LM, Walters CL (1976) Gastric juice nitrite: a risk factor for cancer of the hypochlorhydric stomach? Lancet ii:1037-1039

Sander J, and Seif F (1969) Bakterielle Reduction von Nitrat im Magen des Menschen als Ursache einer Nitrosamin-Bildung, rzneimittel-Forsch. 19:1091-1093

Sandler DP, Everson RB, Wilcox AJ, Browder JP (1985) Cancer risk in adulthood from early life exposure to parents' smoking. Am J Publ Health 75:487-492

Schlag P, Ulrich H, Merkle P, Bockler R, Peter M, Hertarth C (1980) Are nitrite and N-nitroso compounds in gastric juice risk factors for carcinoma in the operated stomach? Lancet i:727-729

Schmahl D, Scherf HR (1983): Carcinogenicity activity of N-nitrosodiethylamine in snakes. Naturwissenschaften. 70:94

Spiegelhalder B (1984) Occupational exposure to N-nitrosamines. Air measurements and biological monitoring. In: O'Neill IK, Von Borstel RC, Miller CT, Long J, Bartsch H (eds) N-nitroso compounds: Occurrence, biological effects and relevance to human cancer. IARC Scientific Publications No. 57, International Agency for Research on Cancer. Lyon, p 937

Spiegelhalder B, Muller J, Drasche H, Preussmann R (1987) N-nitrosodiethanolamine excretion in metal grinders. In: Bartsch H, O'Neill IK, Schulte-Hermann R (eds) The relevance of N-nitroso compounds to human cancer exposures and mechanisms. IARC Scientific Publications No. 84, International Agency for Research on Cancer. Lyon, p 550

Srianujata S, Tonbuth S, Bunyaratvej S, Valyasevi A, Promvanit N, Chaivatsagul W (1987) High urinary excretion of nitrate and N-nitrosoproline in Opisthorchiasis subjects. In: Bartsch H, O'Neill IK, Schulte-Hermann R (eds) The relevance of N-nitroso compounds to human cancer exposures and mechanisms. IARC Scientific Publications no 84, International Agency for Research on Cancer. Lyon, p 544

Stalsburg H, Taksdal S (1971) Stomach cancer following surgery for benign conditions. Lancet ii:1175-1178

Stjernfeldt M, Lindsten J, Berglund K, Ludvigsson J (1986) Maternal smoking during pregnancy and risk of childhood cancer. Lancet i:1350-1352

Sugimura T, Fujimura (1967) Tumor production in glandular stomach of rat by N-methyl-N'-nitro-N-nitrosoguanidine. Nature 216:943

Suzuki K, Mitsuoka T (1984) N-nitrosamine formation by intestinal bacteria. In: O' Neill IK, Von Borstel RC, Miller CT, Long J, Bartsch H (eds) N-nitroso compounds: Occurrence, biological effects and relevance to human cancer. IARC Scientific Publications No. 57, International Agency for Research on Cancer. Lyon, p 263

Tsuda M, Nagai A, Suzuki H, Hayashi T, Ikeda M, Kuratsune M, Sato S, Sugimura T (1987) Effect of cigarette smoking and Dietary factors on the amounts of N-nitrosothiazolidine 4-carboxylic acid and N-nitroso-2-methyl-thiazolidine 4-carboxylic acid in human urine: A correlation study on urinary excretion. In: Bartsch H, O'Neill IK, Schulte-Hermann R (eds) The relevance of N-nitroso compounds to human cancer exposures and mechanisms. IARC Scientific Publications No. 84, International Agency for Research on Cancer. Lyon, p 446

Wagner DA, Shuker DEG, Bilmazes C, Obiedzinski M, Young VR, Tannenbaum SR (1984) Modulation of endogenous synthesis of N-nitrosamino acids in human. In: O' Neill IK, Von Borstel RC, Miller CT, Long J, Bartsch H (eds) N-nitroso compounds: Occurrence, nitrite-preserved meat prepared at home (but not in factories) may be associated with stomach cancer. Nitrosamines have been found in metal-working (cutting) fluids (up to 30,000 ppm) and (at lower levels) in cosmetics. At one time certain herbicides contained low levels of nitrosamines, but these levels have now been reduced.

3.2 The Role of Vitamin C

Vitamin C (ascorbic acid) inhibits nitrosamine formation by reducing nitrous acid to give dehydroascorbic acid and nitric oxide (NO) (Mirvish et al., 1972). Because vitamin C competes with amines for nitrous acid, nitrosamine formation is reduced. Inhibitions of 95 to 99% can be achieved in chemical systems. Other compounds (e.g., urea) react similarly with nitrite and are equally effective at pH 1, but at pH 3 to 5 vitamin C is much more effective than urea. The reason is that ascorbate, which is produced from ascorbic acid around pH 4.3 (its pK_a), reacts with nitrite 240 times more rapidly than does ascorbic acid. Before 1975, 150 ppm nitrite was added to cured meat, and fried bacon contained 20 to 100 ppb nitrosamines. Since 1975, only 120 ppm nitrite is allowed in all nitrite-preserved products in this country, and these must also contain 500 ppm ascorbate or its isomer, erythorbate. As a result, nitrosamines in fried bacon have been reduced to 0 to 20 ppb.

Consumption of fresh fruits and vegetables, especially those with high vitamin C levels, is negatively correlated with the incidence of cancer of the stomach, esophagus, mouth, larynx and cervix. This negative correlation suggests that these cancers are caused by N-nitroso or other nitrite-derived compounds and that their formation is inhibited by vitamin C. Most of these foods also contain polyphenols which react with nitrite and might act similarly with vitamin C. Ascorbic acid is water-soluble, and the reaction with nitrite must take place in water. Vitamin E (α-tocopherol) is a lipid-soluble phenol which reacts with nitrite to produce NO and a quinonoid derivative, and can destroy nitrous acid or nitrogen oxides in lipidic media. Vitamin E inhibited by 99% the nitrosation of N-butylacetamide by N_2O_4 (the dimer of NO_2) in organic solvents (Mirvish, 1981).

3.3 Origin of Gastric Nitrite

In the United States, the average human stomach receives 4.6 mg nitrite per day from the mouth, and the fasting stomach contains 120 μg nitrite per liter. Dietary nitrite, of which

40% arises from nitrite-cured meat, supplies 20% (0.9 mg/day) and nitrite in saliva supplies 80% of gastric nitrite. An unknown amount of gastric nitrate is reduced to nitrite by bacteria, especially if the stomach has a high pH. Dietary nitrate, which is the origin of saliva nitrite, is normally 75 mg/day in the United States, and 90% of this comes from vegetables. Water usually contains little nitrate, but nitrate intake from drinking water is 75 mg/day, assuming an intake of 1.5 liters per day of water with 50 ppm nitrate. Under these conditions nearly half the nitrate intake comes from vegetables, but they are protective for certain cancers because (in my opinion) the increased gastric nitrite arising from the nitrate is counter-balanced by the ascorbate and polyphenols contained in these foods. This does not apply to nitrate in drinking water. Dietary nitrate is absorbed in the stomach and small intestine, and travels via the blood to the salivary glands. In these glands, an active transport system for anions transports nitrate (and also thiocyanate) into the saliva, which contains nitrate but not nitrite when collected directly from the salivary glands. Bacteria in the mouth reduce nitrate to nitrite, which is swallowed in the saliva. This reduction totals 5% of dietary nitrate.

3.4 *In vivo* Formation of N-nitroso Compounds

Sander and Bürkle (1969) discovered that rodents fed amines and nitrite can produce nitrosamines and develop nitrosamine-induced tumors. Similar findings were reported in our 1971 study in which four amines were given in the food and sodium nitrite was given in the drinking water (Mirvish, 1975). Lung adenomas (benign tumors) were induced with three of these amines and were attributed to nitrosamine formation in the stomach; N-Alkylureas and nitrite produced similar results. The experiment with amines worked when weakly basic amines, which react readily with nitrite, were used. No tumors developed when the strongly basic dimethylamine was given with nitrite. The number of tumors per mouse was proportional to the dose of amine (piperazine) at a constant nitrite dose and increased in proportion to nitrite concentration squared when piperazine was kept constant. This evidence accords with the kinetics discussed above and indicates that *in vivo* nitrosamine formation (nitrosation) is simply an acid-catalyzed reaction occurring in the stomach. The quadratic curve obtained for tumor yield versus nitrite means that the number of tumors increased rapidly above a level of about 0.5 g nitrite/L water; that is, the tumor yield at low nitrite doses was less than that predicted in humans if the response had been linear. This concept is important for evaluating the toxic effects of low nitrate levels.

Giving vitamin C with amines in the diet inhibited tumor induction when nitrite was given in drinking water, and inhibitions of up to 89% were achieved with 2.3% ascorbate in the diet (Mirvish, 1975). Nitrosation of morpholine was inhibited more effectively than that of piperazine, which reacts more rapidly with nitrite. The inhibition worked particularly well for methylurea plus nitrite. In later studies, nitrosamine formation was measured directly by chemical analysis. For example, Iqbal et al. (1980) detected 940 ng N-nitrosomorpholine per mouse in mice killed after treatment with morpholine and nitrite. When ascorbate also was given, nitrosomorpholine production was inhibited by 97%.

In closed aerobic systems, where NO produced from the ascorbate-nitrite reaction is not rapidly removed, the NO can be oxidized by oxygen to NO_2, which reacts with water to regenerate half the original nitrite. In such systems, therefore, twice as much ascorbate is needed as in open or anerobic systems. Factors controlling ascorbate inhibition of gastric nitrosamine formation include rates of mass transfer of nitrogen oxides between liquid and gas phases, and the rate of oxygen transfer from blood to stomach contents (Licht and Dean, 1988; Licht et al., 1988).

$$2 \ HNO_2 + ascorbate \rightarrow 2 \ NO + dehydroascorbate$$
$$2 \ NO + O_2 \rightarrow 2 \ NO_2 \ (in \ gas \ or \ liquid \ phase)$$
$$2 \ NO_2 + H_2O \rightarrow HNO_2 + HNO_3$$

Nitrosating agents were produced in the skin lipids of mice exposed to atmospheric NO_2. The principal component of the lipids was cholesterol nitrite (Mirvish et al., 1986). This nitrite ester can react with amines *in vitro* and probably also *in vivo* to produce nitrosamines. Vitamins C and E (especially vitamin E) also can inhibit tumor production by preformed nitrosamines and other carcinogens (Mirvish, 1981, 1986), and vitamin C or E inhibition of tumorigenesis by amines and nitrite could be partly due to this effect.

3.5 *In vivo* Production of Nitrate and Nitrosamines by Macrophages and Bacteria

Thus far we have assumed that nitrosamines are produced only in the stomach by acid-catalyzed chemical nitrosation, and that nitrite arises only from exogenous nitrate exposure. However, activated macrophages (part of the immune system) synthesize NO, which can be converted to nitrite, nitrate and nitrosamines, and certain bacteria also can catalyze nitrosation. These subjects are reviewed by J. Hotchkiss (this volume).

4 Epidemological Experience

The cancers most likely to be caused by N-nitroso or other nitrate-derived compounds formed in the body are stomach and esophageal cancer. The following is the main evidence for stomach cancer (Mirvish, 1983). Average nitrate intake in 12 countries correlates with stomach cancer incidence. For example, Japan has three times the U.S. level of nitrate intake (apparently because of extensive reuse of sewage) and seven times the U.S. rate of stomach cancer. Rats fed nitrosamides, especially nitrosoureas and methylnitrosonitroguanidine, develop stomach cancer of a similar type (adenocarcinoma) to that found in humans. Unstable carcinogens like nitrosamides, diazoquinones and monoalkylnitrosamines are most likely to act in the stomach if they are produced there. Certain nitrosatable amines and amides occur in foods associated with stomach cancer, especially dried or smoked fish and vegetables. The negative association of this cancer with fresh fruits and vegetables containing vitamin C has already been mentioned. A high starch diet, which has been linked with stomach cancer, causes the rat stomach to be 1 pH unit more acidic than a high-protein diet, and this would favor nitrosamide formation. In the N-nitrosoproline (NPRO) test, elevated levels of urinary nitrate (indicating increased nitrate intake) and NPRO were found in a high gastric cancer area of Japan (Kamiyama et al., 1987). Nitrite produces a highly mutagenic α-hydroxynitrosamine from a chloroindole in fava beans eaten in a high-gastric-cancer area of Columbia (Buchi et al., 1986). An important cofactor for stomach cancer is highly salted food, common in Japan.

Stomach cancer was the principal cancer in the United States in the 1930s, but its incidence has fallen steadily and it is now only the ninth most common tumor in this country. This success story is not due to cancer scientists! If the nitrite-derived-substance theory is true, this reduction in incidence can be linked to the development of new ways of processing and handling foods (Mirvish, 1983). Introduction of refrigerators in the 1940s prevented nitrite accumulation which had occurred as a result of bacterial reduction of nitrate in soups and food products stored at room temperature. Refrigeration also changed the way foods were preserved. Previously, farmers preserved pork or beef with handfuls of nitrate salts and common salt, and the products were widely consumed because fresh meat could not be stored. Consumers were consequently exposed to high nitrate and salt levels. Nitrite-preserved meat is now prepared in large facilities under controlled conditions, and more meat is eaten fresh.

In addition, fresh and frozen fruits and vegetables are now supplied year-round by a national and international distribution system.

Esophageal cancer in Western countries is probably due to nitrosamines in tobacco, which is a risk factor together with alcohol in the development of this cancer (Hecht and Hoffmann, 1989). The incidence of esophageal cancer in Western countries, however, is relatively low. In contrast, esophageal cancer shows very high rates among blacks of Transkei, South Africa; among Turkomanis of Iran; and in Linxian county, China (Day, 1984). In these areas, esophageal cancer could be due to gastric nitrosamine formation. Evidence for this causation includes the following. Certain nitrosamines (and no other compounds) induce esophageal cancer in rats (Preussmann and Stewart, 1984). Excess nitrate and vitamin C deficiency have been associated with esophageal cancer in these areas (National Academy of Sciences, 1981). O^6-Methylguanine, which can be produced by nitrosamines, was detected in the esophageal DNA of esophageal-cancer patients (Umbenhauer et al., 1985). Human esophagus was reported to form the DNA adduct N^7-methylguanine from dimethylnitrosamine (Harris et al., 1979; Autrup and Stoner, 1982); and human esophageal microsomes produce formaldehyde from the rat esophageal carcinogen, methylamylnitrosamine, indicating that they can activate this nitrosamine (Huang et al., 1990). Finally, the NPRO test showed elevated levels of urinary NPRO and nitrate in a high-esophageal-cancer area of China (Lu et al., 1986).

Certain populations show high rates of stomach cancer and others show high rates of esophageal cancer, even though both have high nitrate intakes. Among the reasons for this variation could be (1) stomach cancer is due to nitrosamides and esophageal cancer to nitrosamines, each formed *in vivo* from precursors in different foods; and (2) some populations are exposed to gastric cocarcinogens (agents that aid the action of carcinogens), such as high salt, and others are exposed to esophageal cocarcinogens, such as alcohol.

5 The NPRO Test

5.1 Principles

Ohshima and Bartsch (1981) demonstrated that nitrosamines can be produced in humans. A volunteer drank 325 mg nitrate and, 30 minutes later, 500 mg L-proline. This dietary amino acid is a secondary amine and can be nitrosated. The 24-h urine was analyzed by conversion of NPRO to its methyl ester with diazonmethane, followed by gas chromatography and thermal-energy analysis. The urine contained 23 μg NPRO. The NPRO yield was proportional to the proline dose when this was varied with a fixed dose of 325 mg nitrate. When the proline dose was fixed at 500 mg and the nitrate dose was varied, the NPRO yield was proportional to nitrate dose squared, in accord with the chemical kinetics. Ascorbic acid (1 g) inhibited NPRO formation by 81%.

These results indicate the potential for forming carcinogenic nitrosamines. Feeding both proline and nitrate permits one to measure factors in the stomach, other than the supply of amines and nitrate, that affect nitrosation. If only proline is given, the test also includes the contribution of nitrate/nitrite.

NPRO is formed in the stomach, absorbed into the blood and excreted in the urine. The test is considered safe because NPRO produced no tumors in rats when up to 20 g per animal was given. Little if any NPRO is metabolized by rats. Because about 1 μg NPRO is normally excreted, increasing this amount for a few days is unlikely to be hazardous. Elevated NPRO formation was found in individuals who were cigarette smokers (Bartsch et al., 1984; Hoffmann and Brunnemann, 1983; Ladd et al., 1984) and in areas of East Asia with high stomach or esophageal cancer (see above). Administration of vitamin C lowered NPRO formation in all these studies, demonstrating that NPRO was produced in the body and not consumed in the diet.

5.2 Selected Results from this Laboratory

NPRO excretion was measured in volunteers taking nitrate and proline with standard meals or living in rural Nebraska and drinking high- or low-nitrate water. This project was undertaken in collaboration with A.C. Grandjean and S. Fike (Center for Human Nutrition, Omaha); and T. Maynard, L. Jones, S. Rosinsky and G. Nie (Eppley Institute for Research in Cancer and Allied Diseases). Our first study was designed to examine conditions in the stomach that affect NPRO formation; the second was designed to examine the effects of drinking high- or low-nitrate water. In the first study, adult male subjects followed a low-NPRO diet for 5 days. On days 4 and 5, chewing gum and high-ascorbate foods were not allowed and a standard 650-calorie test lunch was eaten. L-Proline (500 mg) was given with this meal and sodium nitrate (400 mg nitrate) was given 5 min to 2 h before the meal. Urine (24-h) and saliva samples were collected. The urine samples were analyzed for NPRO, nitrate, specific gravity and creatinine. The saliva samples were analyzed for nitrite and nitrate. With nitrate given just before and proline given with the meal, mean NPRO was 2.1 μg, significantly more than the 0.6 μg in control subjects who received only the standard meal. Giving 1 g ascorbic acid with the proline and nitrate reduced the mean NPRO to 1.2 μg. The NPRO yield increased to 11.1 μg when nitrate and proline were taken while fasting, demonstrating that the presence of food in the stomach inhibits nitrosation. Mean NPRO yield was 2.1, 4.9 and 2.3 μg when nitrate was taken 5 min, 1 h and 2 h before the meal, consistent with a model in which NPRO is formed for 1 h after taking proline. Therefore, people should not take nitrosatable compounds (for example, certain drugs) while fasting. Drinking nitrate-rich water could be most hazardous 1 h before a meal.

In the second study, similar NPRO tests were performed by 44 rural Nebraska men with high- or low-nitrate drinking water from private wells. The 5-day test included the same regimen as before, urine collection on day 4 while the subjects followed their usual activities and ate near-normal diets, and urine collection on day 5 after an overnight fast, drinking 2 glasses of well water and taking 500 mg proline at 6 and 7 a.m., and continued fasting until 1 p.m. When there was no nitrate in the drinking water, basal NPRO was 1 μg/day. The 15 highest NPRO values were 3 to 6 μg/day. On day 4 the 18 men drinking well water with more than 18 ppm nitrate-N showed a mean of 2.8 μg NPRO, significantly more than the 1.4 μg for the 26 men drinking water with less than 18 ppm nitrate-N, but there was no significant difference on day 5. The high-nitrate group excreted more urine nitrate than the low-nitrate group on both days. A comparison of parameters for all results on each day revealed significant ($P < 0.05$) correlations for urine NPRO and urine nitrate versus water nitrate on both days, for saliva nitrite and saliva nitrate versus water nitrate on day 5, and for urine creatinine versus urine NPRO and urine nitrate on day 4. Hence, increased water nitrate levels led to increased excretion of NPRO. Although the number of subjects was small, it appears that urine NPRO began to increase above 30 ppm nitrate-N in the water. These findings are similar to those of Moller et al. (1989).

6 Conclusions

People are exposed to N-nitroso compounds as such and via their formation in the body from nitrite and, indirectly, nitrate. The relevance of this exposure to cancer causation seems clear for tobacco products containing preformed nitrosamines and is likely for cancers that may be due to *in vivo* formation of N-nitroso compounds. The NPRO test is useful for evaluating the risk of this *in vivo* formation.

Acknowledgements

Our research was supported by National Institutes of Health grants RO1-CA-35628 and UO1-CA-43236, core grant CA-36727 from the National Cancer Institute, core grant SIG-16 from the American Cancer Society, and a grant from the American Institute of Cancer Research.

References

Autrup H, Stoner GD (1982) Metabolism of N-nitrosamines by cultured human and rat esophagus. Cancer Res 42:1307-1311

Bartsch H, Ohshima H, Munoz N (1984) *In vitro* nitrosation, precancerous lesions and cancers of the gastrointestianal tract: On-going studies and preliminary results. In: O'Neill IK, Von Borstel RC, Miller CT, Long J, Bartsch H (eds.) N-Nitroso Compounds: Occurrence, Biological Effects and Relevance to Human Cancer. IARC Sci Publ 57:957-964. Int. Agency Res. Cancer, Lyon

Buchi G, Lee GCM, Yang D, Tannenbaum SR (1986) Direct acting, highly mutagenic α-hydroxynitrosamines from 4-chloroindoles. J Am Chem Soc 108:4115-4119

Conboy JJ, Hotchkiss JH (1989) A photolytic interface for HPLC: Chemiluminescence detection of nonvolatile N-nitroso compounds. Analyst 114:155-159

Day NE (1984) The geographic pathology of cancer of the oesophagus. Brit Med Bull 40:329-334

Foreman D, Shuker DED (eds.) (1989) Nitrate, nitrite and nitroso compounds in human cancer. Cancer Surveys 8:205-487

Harris CC, Autrup H, Stoner GD, Trump BF, Hillman E, Schafer PW, Jeffrey AM (1979) Metabolism of benzo(a)pyrene, N-nitrosodimethylamine, and N-nitrosopyrrolidine and identification of the major carcinogen-DNA adducts formed in cultured human esophagus. Cancer Res 39:4401-4406

Hecht S, Hoffmann D (1989) The relavance of tobacco-specific nitrosamines to human cancer. Cancer Surveys 8:273-294

Hoffmann D, Brunnemann KD (1983) Endogenous formation of N-nitrosoproline in cigarette smokers. Cancer Res 43:5570-5574

Huang Q, Mirvish SS, Stoner G, Nickols J (1990) Metabolism of methyl-n-amylnitrosamine by human esophagus. Proc Am Ass Cancer Res 31:117

Iqbal ZM, Epstein SS, Krull IS, Goff U, Mills K, Fine DH (1980) Kinetics of nitrosamine formation in mice following oral administration of trace-level precursors. In: Walker, EA, Gricuite L, Castegnaro M, Borzsonyi M (eds.) N-Nitroso Compounds: Formation and Occurrence. Sci Public Int Agency Res Cancer, Lyon, p 169

Kamiyama S, Ohshima H, Shimada A, Saito N, Bourgade MC, Ziegler P, Bartsch H (1987)

Urinazry excretion of N-nitrosamino acids and nitrate by inhabitants in high- and low-risk areas for stomach cancer in Northern Japan. In: Bartsch H, O'Neill I, Schulte-Hermann R (eds.) The Relevance of N-Nitroso Compunds to Human Cancer: Exposures and Mechanisms, IARC Sci Publ 84, Int Agency Res Cancer, Lyon, p 497

Ladd KF, Newmark HL, Archer MC (1984) N-Nitrosation of proline in smokers and nonsmokers. J Natl Cancer Inst 73:83-87

Licht RL, Tannenbaum SR, Dean WM (1988) Use of ascorbic acid to inhibt nitrosation: kinetic and mass transfer considerations for an *in vitro* system. Carcinogenesis 9:365-372

Licht RL, Dean WM (1988) Theoretical model for predicting rates of nitrosamine and nitrosamide formation in the human stomach. Carcinogenesis 9:2227-2237

Lu SH, Ohshima H, Fu HM, Tian Y, Li FM, Blettner M, Wahrendorf J, Bartsch H (1986) Urinary excretion of N-nitrosamino acids and nitrate by inhabitants of high- and low-risk areas for esophageal cancer in Northern China: Endogenous formation of nitrosoproline and its inhibition by vitamin C. Cancer Res 46:1485-1491

Mirvish SS (1975) Formation of N-Nitroso compounds: Chemistry, kinetics and *in vivo* occurrence. Toxico Applied Pharmacol 31:325-351

Mirvish SS (1981) Inhibition of the formation of carcinogenic N-nitroso compounds by ascorbic acid and other compounds. In: Burchenal JH, Oettgen HP (eds.) Achievements, Challenges. Prospects for the 1980s. Grune and Stratton, New York, p 557

Mirvish SS (1983) The etiology of gastric cancer: Intragastric nitrosamide formation and other theories. J Natl Cancer Inst 71:631-647

Mirvish SS (1986) Effects of vitamins C and E on N-nitroso compound formation, carcinogenesis, and cancer. Cancer 58:1842-1850

Mirvish SS, Wallcave L, Eagen M, Shubik P (1972) Ascorbate-nitrite reaction: possible means of blocking the formation of carcinogenic N-nitroso compounds. Science 177: 65-68.

Mirvish SS, Babcook DM, Deshpande AD, Nagel DL (1986) Identification of cholesterol as a mouse skin lipid that reacts with nitrogen dioxide to yield a nitrosating agent (NSA), and of cholesteryl nitrite as the NSA produced in a chemical system from cholesterol. Cancer Lett 31:97-104

Moller H, Landt J, Pedersen E, Jensen P, Autrup H, Jensen OM (1989) Endogenous nitrosation in relation to nitrate exposure from drinking water and diet in a Danish rural population. Cancer Res 49:3117-3121

National Academy of Sciences (1981) The Health Effects of Nitrate, Nitrite and N-Nitroso Compounds. National Academy Press, Washington, D.C.

Ning JP, Yu MC, Wang QS, Hendrson BE (1990) Consumption of salted fish and other risk factors for nasopharyngeal carcinoma (NPC) in Tianjin, a low-risk region for NPC in the People's Republic of China. J Natl Cancer Inst 82:291-296

Ohshima H, Bartsch H (1981) Quantitative estimation of endogenous nitrosation in humans by monitoring N-nitrosoproline excreted in the urine. Cancer Res 41:3658-3662

Preussmann R, Stewart BW (1984) N-Nitroso carcinogens In: C.E. Searle (ed.) Chemical Carcinogens. Am Chem Soc 2:643-828.

Sander J, Bürkle G (1969) Induktion maligner tumoren bei ratten durch gleichzeitige verfutterung von nitrit und sekundaren aminen. z. Krebsforsch 73:54-66

Shephard S, Lutz W (1989) Nitrosation of dietary precursory. Cancer Surveys 8:401-421

Umbenhauer D, Wild CP, Montesano R, Saffhill R, Boyle JM, Huh N, Kirstein U, Thomale J, Rajewsky MF, Lu SH (1985) O^6-Methyldeoxyguanosine in oesophageal DNA among individuals at high risk of oesophageal cancer. Int J Cancer 36:661-665

Wakabayashi K, Nagoo M, Singamura T (1989) Mutagens and carcinogens produced by the reaction of environmental aromatic compounds with nitrite. Cancer Surveys 8:385-399

Walton G (1951) Survey of literature relating to infant methemoglobinemia due to nitrate-contaminated water. Am J Public Health 41:988-996

EPIDEMIOLOGICAL STUDIES OF THE ENDOGENOUS FORMATION OF N-NITROSO COMPOUNDS

H. Møller and D. Forman[1]
Danish Cancer Registry, Institute of Cancer Epidemiology,
Danish Cancer Society, Rosenvængets Hovedvej 35, PO Box 839,
2100 Copenhagen Ø., Denmark (HM)

Abstract

Carcinogenic N-nitroso compounds may be formed endogenously in the stomach. The rates of formation of nitrosoproline in human subjects increase with nitrate intake and decrease with ascorbate intake. The epidemiologic evidence linking endogenously formed N-nitroso compounds and cancer risk in man is still inconclusive.

1 Introduction

N-nitroso compounds (NOC) have been recognized as carcinogenic since the pioneering work of Magee and Barnes (1956). The discovery that NOC may form endogenously, and, indeed, the finding that such endogenous formation of potential carcinogens in humans may depend on nitrate intake (Ohshima and Bartsch, 1981), has increased awareness of the potential health hazards of high-nitrate intake. The most serious concern has been the possible association with gastric-cancer risk (Crespi, this volume; Mirvish, this volume; Correa et al., 1975; Correa, 1983, 1988).

Development of the N-nitrosoproline test (the NPRO test) (Ohshima and Bartsch, 1981) opened an entirely new direction for epidemiological studies of endogenous formation of NOC and the determinants of gastric-cancer risk. The amino acid proline is a constituent of the normal diet; especially under acidic conditions, proline is nitrosated to form N-nitrosoproline (NPRO). NPRO is not mutagenic or carcinogenic, and the compound is not metabolized but is excreted quantitatively in the urine. NPRO excretion, most often measured after an oral dose of proline, can thus act as a noninvasive biologic marker of the level of endogenous formation of NOC in the stomach (Mirvish, this volume).

---

[1] ICRF Cancer Epidemiology Unit, Radcliffe Infirmary, Oxford OX2 6HE, UK (DF)

NATO ASI Series, Vol. G 30
Nitrate Contamination
Edited by I. Bogárdi and R. D. Kuzelka
© Springer-Verlag Berlin Heidelberg 1991

Satisfactory epidemiological studies of nitrate intake and cancer risk have been very difficult to design for two reasons. First, the relevant period of nitrate exposure may be many decades prior to the onset of disease. Second, it is now clear that a complex network of causation is involved in the carcinogenic process. The NPRO test offers a new possibility for solving this second design problem. The process of endogenous formation of carcinogens, and the action of such compounds on the cells of the gastric mucosa and elsewhere in the body, are each processes which may be influenced by still other factors (for example, tobacco smoking, intake of nitrosation inhibitors, intake of free radical scavengers, and the types and amounts of nitrosable compounds in the diet) (Forman, 1987, 1989; Howson et al., 1986). The use of biologic markers such as NPRO excretion as determinants of disease risk or as endpoints in epidemiological studies may help to clarify these relationships. The NPRO test also is significant in that it maintains the primary quality of epidemiological research in comparison with laboratory experimentation: the study of disease occurrence and causation in a statistically representative human population, under conditions which resemble the normal situation in the population under study.

In the following sections we shall review epidemiological studies of NPRO formation in relation to cancer risk and epidemiological studies of the determinants of NPRO excretion itself.

2 Epidemiological Studies of Cancer Risk in Relation to NPRO Excretion

The first study examined NPRO formation in two areas of China with different rates of oesophageal cancer, and involved 84 subjects (Lu et al., 1986). NPRO was measured in 24-h urine collected (1) during a period in which the subjects ingested 3 x 100 mg proline (test-NPRO), (2) during a period in which the subjects ingested 3 x 100 mg proline and 3 x 100 mg ascorbate (ascorbate-NPRO), and (3) during a period with no loading dose of proline or ascorbate apart from amounts present in the ordinary diet, which was unrestricted in most of these studies (background-NPRO).

Background-NPRO was higher in the high-risk area than in the low-risk area (5.7 vs. 1.7 μg NPRO/24 h) (Table 1). This difference may be due to a higher dietary intake of proline, or indeed, of preformed NPRO, in the high-risk area. The urinary NPRO excretion after a large

Table 1. Epidemiologic studies involving comparison of urinary NPRO excretion in populations of low and high cancer risk

Study Population (n)		Relative risk of cancer[1] in men	Background-NPRO	Test-NPRO	Ascorbate NPRO	Nitrate Exposure	Reference
			μg/24 h	μg/24 h	μg/24 h	mg/24 h	
China							
Linxian	(44)	4.3	5.7	8.3	2.4	94	Lu et al., 1986
Fanxian	(40)	1.0	1.7	4.4	ND	48	
			p<0.001	p<0.001	p<0.05	p<0.001	
Japan							
Akita	(52)	3.1	3.8	12.6	3.2	95	Kamiyama et al., 1987
Iwate	(52)	1.0	6.1	7.1	4.9	145	
			NS	p<0.001	p<0.05	p<0.001	
			μg/12 h	μg/12 h	μg/12 h	mg/12 h	
Poland							
Rural	(43)	2.1	1.8	2.8	1.2	106	Zatonski et al., 1989
Urban	(45)	1.0	2.0	2.4	1.8	76	
			NS	NS	NS	p<0.001	
			μg/12 h	μg/12 h	μg/12 h	mmol/12 h	
Italy							
Florence	(40)	2.9	0.94	1.59	ND	37	Knight et al., 1990
Cagliari	(40)	1.0	0.58	1.44	ND	34	
			p<0.05	NS		NS	
			μg/12 h	μg/12 h	μg/12 h	mmol/12 h	
Costa Rica							
Turrubares	(26)	3.2	ND	0.84	0.66	0.23	Sierra et al., 1990
Hojancha	(25)	1.0	ND	0.54	0.28	0.20	
				p<0.04	p<.002	NS	

ND: not determined; NS: not significant
[1] China: oesophageal cancer; other studies: gastric cancer.

loading dose of proline (test-NPRO, which measures more directly the chemical activity of nitrosating agents in the stomach) was higher in the high-risk area than in the low-risk area (8.3 vs. 4.4 μg NPRO/24 h). When ascorbate was ingested with the loading dose of proline in the high-risk population, urinary NPRO dropped to a level even below the background level (2.4 μg NPRO/24 h). Urinary nitrate excretion was substantially higher in the high-risk area than in the low-risk area (94 versus 48 mg/24 h).

The results of this study seem to fit very nicely with theoretical expectations: the higher intake of nitrate in the high-risk population leads to a higher rate of endogenous nitrosation of nitrosable agents in the diet (as indicated by the increased NPRO excretion in the test situation), and may, in part, explain the higher risk of oesophageal cancer in this population. The clear inhibitory effect of ascorbate suggests a potential for primary prevention.

This type of analysis, however, has some major limitations. First, it does not account for other important differences between the two geographic populations which alone may explain the difference in cancer risk. Second, it does not account for the latency period of cancer development; cancer incidence or mortality may occur several decades after exposure.

In a similar Japanese study (Kamiyama et al., 1987), the population with high mortality from gastric cancer again had higher test-NPRO levels than the low-risk population, but nitrate excretion was higher (although not statistically significant) in the low-risk population (Table 1).

In a Polish study (Zatonski et al., 1989), nitrate excretion increased significantly in the high-risk population, but there was no significant difference in test-NPRO levels. NPRO excretion was much lower in the Polish subjects than in the Japanese and Chinese subjects (Table 1).

The substantial natural variation in gastric-cancer risk in different areas in Italy has similarly been used for comparison of NPRO excretion in low- and high-risk populations (Knight et al., 1990). No significant difference in test-NPRO or in urinary nitrate was found, but background-NPRO was highest in the high-risk area (Table 1).

It has been suggested that gastric cancer is a process which is initiated very early in life, and that endogenous nitrosation in childhood may be critical in determining the future cancer risk.

A study comparing NPRO excretion in Costa Rican schoolchildren in two areas with different gastric cancer incidence showed no difference in nitrate excretion, but a significantly higher test-NPRO was found in the high-risk area than in the low-risk area (Sierra et al., 1990) (Table 1).

Considering these studies together, it appears that test-NPRO in all studies is highest in the high-risk area but most clearly so (and only with statistical significance) in China and Japan.

A correlation study conducted in China using pooled urinary samples and cancer mortality data from 26 counties shows a weak, positive correlation between test-NPRO and oesophageal cancer mortality ($r=0.30$) but a negative correlation between test-NPRO and gastric cancer mortality ($r=-0.28$) (Chen et al., 1987). Researchers also found a significant inverse correlation between plasma ascorbate levels and the risk of oesophageal cancer. From the studies listed in Table 1, it appears consistently to be the case that ascorbate taken together with the proline dose effectively inhibits excretion of NPRO.

3 Determinants of NPRO Formation

Recent studies have attempted to use the NPRO test as a tool for identification of factors which determine the rate of endogenous nitrosation within human populations. The role of waterborne nitrate on gastric-cancer risk has been much debated (for reviews see Fraser, 1985; Forman, 1989). The Danish study of endogenous nitrosation in relation to nitrate concentration in drinking water and total nitrate intake was designed in order to evaluate the health risks imposed by the high and increasing levels of nitrate in drinking water in certain parts of Denmark (Møller et al., 1989a, 1989b). The study makes use of the very unusual situation in northern Jutland where, within an otherwise homogenous community, there is very high variation in nitrate concentration in the drinking water from small, local waterworks.

The importance of drinking water nitrate in relation to overall nitrate intake was quantified. When the nitrate concentration is about 45 mg/L, the total daily nitrate intake in this Jutland population is 89 mg, 58% of which is derived from the drinking water. When the nitrate concentration is about 80 mg/L, the total daily intake is 123 mg, 70% of which is derived from the drinking water. Estimates from two English studies are in close agreement with this data (Chilvers et al., 1984; Royal Commission on Environmental Pollution, 1979). Urinary NPRO excretion (test-NPRO) ranged from less than 0.5 to more than 100 μg/12 h with a median of 0.5 μg 12 h. Thus, the distribution of urinary NPRO is extremely skewed with a tail towards higher values. For this reason, multivariate statistical analysis was carried out with the use of multiple logistic regression, a regression procedure which is very commonly used in epidemiology for analysis of dichotomous outcomes (for example, dead/alive, sick/ well, or, as in this case, positive test-NPRO/negative test-NPRO). The analysis allows binary as well as continuous independent variables to be considered.

Table 2 shows the distribution of positive and negative NPRO measurements in five categories of total nitrate exposure. The cutoff point separating positive from negative test-NPRO in this table is 1 μg NPRO/12 h. In the lowest category of nitrate exposure (0 to 49 mg/ 24 h) the prevalence of having a positive test-NPRO was 33%; one may also express this in terms of the odds of having a positive test at this level of exposure which is 26:54 = 0.48. Either the prevalence rate or the odds of the outcome may be used as the baseline for a relative measure expressing the effect of nitrate intake on NPRO excretion. For the former, this measure is the prevalence-rate ratio (RR); for the latter, it is the odds ratio (OR). With multiple logistic regression odds, ratios may be determined in a statistical analysis where several determinants of risk are considered simultaneously; the procedure yields adjusted estimates of the odds ratio associated with each factor.

Table 2. Prevalence of excretion of 1 μg NPRO or more in overnight urine after 500 mg loading dose of proline, in relation to total nitrate intake in Danish study

Total nitrate intake, mg/24 h	Number of samples		Prevalence of positive test-NPRO	RR	Odds of positive test-NPRO	OR
	Positive ≥ 1 μg NPRO/12 h	Negative ≤ 1 μg NPRO/12h				
0 - 49	26	54	0.33	1.0	0.48	1.0
50 - 99	44	57	0.44	1.3	0.77	1.6
100 - 149	22	30	0.42	1.3	0.73	1.5
150 - 199	12	12	0.50	1.5	1.00	2.1
200 +	18	9	0.67	2.0	2.00	4.2
			trend: p < 0.01		trend: p < 0.01	

Logistic regression analysis of these data identified three strong determinants of test-NPRO: nitrate intake, tobacco smoking, and consumption of cured meat in the period just prior to urine collection. Some interaction between nitrate intake and smoking was observed. The outcome of test-NPRO exceeding 1 μg/12 h was strongly associated with nitrate intake in nonsmokers. Smokers of one to twenty cigarettes per day had increased test-NPRO which, however, was independent of nitrate intake.

Table 3 shows the dependency of test-NPRO on nitrate intake (nonsmokers only), tobacco smoking (nitrate intake fixed at 40 mg/24 h), sex, age, history of gastric disease, and consumption of cured meat. In order to evaluate also the choice of cutoff point separating positive and negative tests, the analysis were conducted three times with cutoff points at 1, 5, and 10 μg test-NPRO/12 h respectively.

Table 3. Logistic regression analysis of overnight NPRO excretion in relation to nitrate intake, tobacco smoking, and other factors in Danish study. Odds ratios (with 95% confidence intervals for having high urinary NPRO excretion. Separate analyses were conducted with cutoff points at 1, 5, and 10 μg NPRO/12 h

	Cutoff point μg NPRO/12 h		
	1	5	10
Total nitrate intake (per 50 mg/ 24 h); nonsmokers only.	1.90 (1.39-2.58)	1.73 (1.22-2.46)	1.90 (1.19-3.04)
Smoking[1], cigarettes per day 1-20	2.20 (1.08-4.47)	2.69 (0.96-7.53)	5.04 (0.96-26.42)
Smoking[1], cigarettes per day 21+	3.67 (0.62-21.73)	4.54 (0.51-40.16)	14.28 (1.12-181.56)
Sex (men vs. women)	1.28 (0.72-2.29)	1.41 (0.63-3.19)	1.67 (0.49-5.64)
Age (35-49 vs. 20-34)	1.31 (0.67-2.56)	0.82 (0.33-2.04)	2.02 (0.39-10.38)
Age (50-64 vs. 20-34)	1.12 (0.57-2.20)	0.79 (0.32-1.97)	3.40 (0.70-16.51)
History of gastric disease (some vs. none)	0.47 (0.21-1.06)	0.61 (0.19-2.00)	1.59 (0.37-6.76)
Eating cured meat (some vs. none)	3.68 (1.27-10.67)	4.63 (1.54-13.87)	9.30 (2.04-42.40)

[1] Total nitrate intake fixed at 40 mg/24 h

Among nonsmokers, nitrate intake is an important determinant of NPRO formation. The odds ratio is fairly independent of the choice of cutoff point, and the confidence interval is very narrow in all three analyses. This means that the effect of nitrate intake in nonsmokers is a general shift in the distribution of NPRO excretion towards higher values.

The effect of smoking is seen to depend somewhat on the choice of cutoff point. The higher the cutoff point, the higher becomes the odds ratio associated with tobacco smoking. Using the higher cutoff points, the estimates for tobacco smoking become less stable as judged by the wider confidence intervals for the estimated odds ratios. Sex and age do not materially influence NPRO excretion in this population. A history of gastric disease (gastritis or ulcer; self-reported) seems to protect against NPRO formation when judged from the analysis using 1 μg/12 h as the cutoff point, but in the other two analyses there is no effect.

Eating cured meat in the period prior to urine collection is associated with a highly increased urinary NPRO, and it seems that cured-meat consumption especially contributes to the upper end of the NPRO distribution. The result presumably reflects the presence of large amounts of preformed NPRO in some cured-meat products (Dunn and Stich, 1984). The odds ratios in Table 3 are not materially altered when cured meat is excluded from the regression models. This means that the effects of nitrate intake and tobacco smoking are not modified by cured-meat consumption. Consumption of beer, which may be a source of preformed NPRO, did not influence NPRO excretion in the present material.

A similar individual-based analysis has been carried out on the basis of the Italian study by Knight et al. (1990). In these data, region of residence and tobacco smoking did not have any effect on NPRO formation and were hence not included the regression analysis. In the Italian study background-NPRO was measured separately for each individual. Background-NPRO and nitrate intake (here measured by urinary nitrate excretion) were the strongest determinants of test-NPRO excretion in the Italian populations.

The odds ratio for nitrate exposure (upper vs. lower tertile) was 7.5 (1.5 to 38.8), and the odds ratio for background NPRO excretion (high vs. low) was 10.4 (2.2 to 49.0). Mutual adjustment of these odds ratios had only a marginal effect on the estimates, which after adjustment were 7.2 (1.3 to 40.8) and 9.6 (1.9 to 47.5), respectively. The close agreement between the crude and the mutually adjusted odds ratios indicates that the effects of nitrate intake and of background NPRO are independent.

Finally, a study of NPRO excretion in individuals was conducted in Columbia (Stillwell et al., 1990). Urinary samples (24 h) were collected from about 165 persons. No loading dose of proline was used in this study. Positive correlations were seen between nitrate intake and NPRO excretion. Measurements of urinary excretion of DNA-adducts (3-methyladenine and 7-methylguanine), which arise from the repair process after chemical reaction between alkylating agents and DNA, were positively correlated with NPRO excretion, and 7-methylguanine (but not 3-methyladenine) was positively correlated with nitrate intake (Table 4).

Table 4. Correlation coefficients between nitrate exposure and urinary excretion of NPRO and DNA-adducts in Columbian study (Stillwell et al., 1990)

	NPRO	3-methyladenine	7-methylguanine
Nitrate	0.43^{***}	0.05	0.18^{*}
NPRO		0.23^{**}	0.16^{*}
3-methyladenine			0.01

* $p<0.005$
** $p<0.01$
*** $p<0.001$

Because no loading dose of proline was used, the variability in proline and NPRO intake between individuals may have contributed substantially to the variation in NPRO excretion. The correlation coefficients in Table 4 would be higher if a loading dose of proline had been used. Similarly, no attempt was made to control for tobacco smoking. This may cause a reduction in the correlation coefficients involving the two DNA-adducts.

4 Discussion

From the epidemiological studies referred to above, there is clear evidence that the rate of endogenous nitrosation of proline is strongly dependent on the intake of nitrate and ascorbate. Nitrate intake enhances the formation of NPRO while ascorbate inhibits its formation.

The finding of a positive correlation between urinary excretion of NPRO and DNA-adducts in the Colombian study further points to the possibility of endogenous formation of NOC being associated with a potential for carcinogenesis. Although it has yet to be demonstrated that there is a real cancer risk resulting from endogenously formed NOC, it would seem prudent to avoid, where possible, situations in which individuals are exposed to very high amounts of nitrate over a period of many years.

In most populations, vegetables are the major source of nitrate. The epidemiological evidence regarding consumption of vegetables is clear: consumption of vegetables reduces the risk of several forms of cancer, especially gastric cancer. Vegetables may exert a protective effect owing to the content of inhibitors of nitrosation (for example, vitamin C and phenolic compounds) (Bartsch et al., 1988).

The evidence relating NPRO excretion and cancer risk is correlational in nature. Correlation studies are not ideal for prediction of the determinants of disease risk to individuals. This is so primarily because the results of such studies may be subject to biases which arise from the presence of other, perhaps unknown, determinants of disease risk which have different prevalence in the studied populations. At face value, the results relating median-NPRO excretion in populations to cancer mortality or morbidity (Table 1) seem to suggest some degree of association, especially in China and Japan. It is important to note, however, that the possible degree of bias is quite sufficient to explain effects of the magnitude observed in the studies in China and Japan as well as to explain the negative findings in other studies.

The results from studies of cancer occurrence in fertilizer workers (Forman, this volume) do not support the hypothesis that nitrate intake is a determinant of cancer risk in Britain. However, the size of the studies is insufficient to support the conclusion that there is no increased cancer risk in fertilizer workers.

The situation at present is thus that we have firm evidence of an association between nitrate intake and the rate of endogenous formation of at least one N-nitroso compound, but the epidemiological evidence relating nitrate intake or indeed endogenously formed NOC to actual cancer risk is inconclusive.

The use of NPRO as a representative of NOC has been of great benefit in elucidating the relevance of endogenous nitrosation in the human situation. It has proven to be a useful marker, and the development of the NPRO-test represents a qualitative advance in the application of biochemical techniques to epidemiological studies. NPRO does, however, have its limitations. In particular the pH dependency for its formation is substantially different from the nitrosamides and certain nitrosamines, and caution may be required in extrapolating from results obtained with NPRO.

Epidemiological assessment of disease risk in relation to endogenous as well as exogenous NOC has been difficult, not least because the concentrations of NOC to which humans are exposed are often very low, and because chemical measurement of many NOC is inherently difficult. Bearing in mind the very high carcinogenic potential of many NOC in laboratory animals, the lack of epidemiological verification of the laboratory findings may be due to insufficient precision in measurement of exposure in humans.

The highest priority at present should be identification of occupational groups with high-nitrate exposure. It is also possible that there may be groups of patients who have undergone long-term therapy for kidney stones and urinary tract infections with high doses of inorganic nitrate salts. It would be desirable to study such heavily exposed groups in different cultural settings. In order to arrive at firm conclusions regarding the magnitude of cancer risk in individuals, such cohorts must be fairly large, comprising thousands of persons, and most importantly, the length of follow-up must be at least 15-20 years after first exposure to high nitrate concentrations. We know from studies on gastric-cancer risk in patients operated for gastric ulcer that this is the period required before gastric tumours can possibly develop (Møller and Toftgaard, 1990).

The finding that background-NPRO is a strong determinant of test-NPRO, independently of nitrate exposure, may be important. A simple explanation for the effect may be a high dietary intake of proline or preformed NPRO in some individuals. The possibility remains, however, that the rate of endogenous nitrosation varies between individuals independently of the intake of nitrate, proline, and NPRO. It would be very interesting to obtain an estimate of this component of the variation in NPRO excretion between individuals.

Total nitrate exposure per unit of time (or the amount formed endogenously, if the dietary amount is known) can be determined in humans (Hotchkiss, 1988a) as follows:

$$NO_3^- \text{ exposure} = \text{dietary } NO_3^- + \text{endogenous } NO_3^- \tag{1}$$

$$\text{excreted } NO_3^- = \text{dietary } NO_3^- + \text{endogenous } NO_3^- - NO_3^- \text{ catabolized} \tag{2}$$

$$\text{catabolized } NO_3^- = (1 - 0.54) \times (\text{dietary } NO_3^- + \text{endogenous } NO_3^-) \tag{3}$$

substituting [3] into [2] and rearranging gives:

$$\text{endogenous } NO_3^- = \text{excreted } NO_3^-/0.54 - \text{dietary } NO_3^- \tag{4}$$

substituting [4] into [1] and ignoring fecal excretion gives:

$$NO_3^- \text{ exposure} = \text{excreted } NO_3^-/0.54 \tag{5}$$

This measurement is important for assessing overall exposure from exogenous versus endogenous sources. Two studies using $[^{15}N]NO_3^-$ have determined the amount of nitrate synthesized in normal humans. Using $[^{15}N]NO_3^-$ and total nitrate-balance studies, Wagner et al. (1983) found that healthy individuals excrete 1.0 mmol of endogenously formed nitrate per day. Using a similar methodology, Leaf et al. (1987) reported endogenous formation to be 1.2 mmol/day (Table 1). As we have pointed out (Hotchkiss, 1988b), the amount of nitrate to which an individual is exposed is the sum of that taken in from dietary sources and that formed endogenously (the nitrate exposure due to oxidation of inhaled NO_x is small in most cases and ignored).

3 Mechanisms of Endogenous Nitrate Formation

Tannenbaum and coworkers serendipitously discovered that acute gastroenteritis will increase the endogenous formation of nitrate in humans. Animals treated with immunostimulants such as lipopolysaccharide (LPS) increased their nitrate excretion by as much as 25-fold (Wagner et al., 1982). The formation of nitrate and nitrite by stimulated macrophages in culture was subsequently shown by Stuehr and Marletta (1985). Work by Hibbs and coworkers has shown that in macrophages the imino nitrogen of arginine is oxidized to nitric oxide (NO) with citrulline being the other product (Hibbs et al., 1987; Iyengar et al., 1987).

There are at least two biological consequences of these findings. First, the formation of nitric oxide may be linked to several important physiological phenomena including cytotoxicity of macrophages (Granger et al., 1988) and the formation of endothelium-derived relaxing factor

(EDRF). Second, endogenously formed oxides of nitrogen may result in the formation of carcinogenic N-nitroso compounds in tissues distant from the stomach (see review by Marletta, 1989). This raises the possibility that processes which lead to immunostimulation, such as infection, also lead to increased endogenous formation of N-nitroso compounds. This latter effect could explain why chronic infections increase the risk of certain cancers.

Recent reports indicate that chronic infections are associated with increased endogenous nitrate synthesis, and in some cases, N-nitroso-compound formation. Tricker et al. (1989) have shown that Schistosomiasis infections result in an increase in urinary nitrite, nitrate and N-nitroso compounds. We have investigated nitrate balance in woodchucks infected by a hepatitis virus. This virus is biologically similar to human hepatitis-B virus and causes hepatic tumors in woodchucks (Popper et al., 1987). Preliminary data show that woodchucks infected with the hepatitis virus endogenously form nearly 6-fold more nitrate than control animals (Table 2).

Table 2. Nitrate balance in woodchucks infected with hepatitis virus; infected vs. uninfected woodchucks (preliminary)

		Day					
		2	3	4	5	$\bar{x}$	Endogenous
Uninfected	dietary intake	2.22[a]	1.16	1.82	3.62	2.42	0.79[b]
	excrete	2.27	1.38	2.21	3.98	2.38	
Infected	dietary intake	2.47	3.51	4.97	5.45	4.26	4.66
	excrete	3.49	5.46	8.99	10.03	6.60	

[a] μmol/day/animal. Average of 4 control and 4 infected animals.

[b] Endogenously synthesized $= \dfrac{\text{excreted } NO_3^-}{0.74} - \text{dietary } NO_3^-$; recovery (24 hr) of an oral dose of $^{15}NO_3^-$ from urine averaged 74%.

4 Nitrate-Mediated Endogenous Formation of N-Nitroso Compounds

The endogenous formation of an N-nitroso compound in humans resulting from the ingestion of nitrate has been amply demonstrated. Ohshima and Bartsch (1981) found increased levels of N-nitrosoproline in human urine after dosing with nitrate and proline. This protocol has since been widely used to study the endogenous formation of N-nitrosoproline (Bartsch et al.,

POTENTIAL HEALTH CONSEQUENCES OF GROUND-WATER CONTAMINATION BY NITRATES IN NEBRASKA

D.D. Weisenburger
Department of Pathology and Microbiology
Eppley Institute for Research in Cancer and Allied Diseases
University of Nebraska Medical Center
Omaha, Nebraska 68198-3135 U.S.A.

Abstract

Ground water serves as the primary source of drinking water for nearly all of rural Nebraska. However, ground-water contamination by nitrates, largely due to the use of fertilizers, is an increasing problem. In an ecologic study, the author found that counties characterized by high fertilizer usage and significant ground-water contamination by nitrates also had a high incidence of non-Hodgkin's lymphoma. Other potential health effects of nitrates in drinking water are also discussed.

1 Introduction

Ground water serves as the primary domestic water supply for approximately 90% of the rural population and 50% of the total population in North America (Power and Schepers, 1989). In Nebraska, nearly all rural households and 84% of public water supplies rely on ground water to meet drinking water needs (Exner and Spalding, 1990). Surveys of large databases in midwestern agricultural states such as Iowa, Kansas, and South Dakota have shown that 25% or more of private wells exceed the drinking-water standard (10 ppm) for nitrate-nitrogen (Halberg, 1988). In Nebraska, 17.5% of domestic wells and 14.2% of public supply wells exceed this standard, and, in some areas, the levels of contamination are increasing at rates of 0.5 to 1.0 ppm per year (Exner and Spalding, 1990). In 1989, 38 community water systems in Nebraska exceeded the nitrate standard. Consequently, public concern regarding the protection of ground water from nitrate contamination is rising.

An increased incidence and mortality due to non-Hodgkin's lymphoma (NHL) has been reported for certain areas of Nebraska (Weisenburger, 1985). Analysis of NHL by histologic type in Nebraska has also revealed an excess of clinically-aggressive forms of NHL (Harrington et al., 1987). Since N-nitroso compounds are known to induce NHL in

NATO ASI Series, Vol. G 30
Nitrate Contamination
Edited by I. Bogárdi and R. D. Kuzelka
© Springer-Verlag Berlin Heidelberg 1991

experimental animals (Mirvish et al., 1987), we decided to perform an ecologic study of the relationship of NHL to various agricultural practices in Nebraska.

2 Methods

All patients diagnosed with NHL, age 21 years or older, and residing in the 66 counties of eastern Nebraska were identified for 1984. The cases were identified through the Nebraska Lymphoma Study Group Registry with the aid of the Nebraska Cancer Registry, area hospitals, and practicing oncologists. Although not an ongoing population-based registry, special procedures were instituted to ascertain all cases in eastern Nebraska. One-hundred and fifty-nine cases of NHL (77 males, 82 females) were identified. The success of these procedures was evidenced by the comparability of the 1984 incidence rate of NHL (age ≥ 21 years) in eastern Nebraska (17.6/100,000) to the expected rate (20.5/100,000) based on age-standardized data from the National Cancer Institute-sponsored Surveillance, Epidemiology, and End Results Registry for 1981-1985 (National Cancer Institute, 1988).

The 66 counties in eastern Nebraska were ranked according to various agricultural factors including bushels of corn produced, acres treated with commercial fertilizers, total pesticides, insecticides for insects on hay and other crops, and herbicides for weeds, grass, or brush in crops and pasture. The data on corn production was obtained from 1981 Nebraska Agricultural Statistics (1983), and 1978 data on fertilizer and pesticide use was obtained from the United States Census of Agriculture (1982). The 66 counties were also ranked according to the percentages of private, municipal, and other water wells that were contaminated by nitrate-nitrogen in excess of 10 ppm. Data on nitrate contamination of well water for 1978-79 was obtained from the Nebraska Department of Health, and from U.S. Geological Surveys of Nebraska (Water Resources Data for Nebraska, 1978-1979).

The 66 counties were then divided into three groups based on their rank with regard to the various agricultural factors (low = 22, intermediate = 21, high = 22). Douglas County, a largely urban area wherein approximately 30% of the population of the 66 counties resides, was deleted from the analysis (47 cases) because the intent of the study was to examine the relationship of agricultural factors to the incidence of NHL. The age-standardized incidences of NHL for the low- and high-risk county groups with regard to each agricultural factor were then calculated and compared. For the well water analysis, however, the incidence of NHL in the 25 counties with less than 10% of the wells contaminated by nitrates (> 10 ppm nitrate-nitrogen) was compared to the incidence in the 25 counties with equal to or greater than 20% of the wells contaminated. The 95% confidence intervals (CI) for the differences between the incidence rates were obtained by using the normal distribution approximation to the binomial.

3 Results

The incidence of NHL in persons age 21 years or older in the 66 counties of eastern Nebraska (17.6/100,000) was not significantly different from that expected (20.5/100,000). Also, the age-standardized incidences of NHL were not significantly increased in county groups characterized by high corn production or large numbers of acres treated with pesticides of all types, insecticides, or herbicides. However, the incidence of NHL in the county group characterized by high commercial fertilizer usage (19.5/100,000) was significantly greater than the incidence in the county group characterized by low fertilizer usage (13.9/100,000; 95% CI = 1.1, 9.0). Similarly, the incidence of NHL in the county group with equal to or greater than 20% of the water wells contaminated by nitrates (21.0/100,000) was significantly greater than in the county group with less than 10% contaminated wells (11.6/100,000; 95% CI = 5.9, 12.8).

4 Discussion

The acute health effects of ground-water contamination by nitrates consist primarily of methemoglobinemia, either clinical or subclinical, occurring in infants under six months of age (Comley, 1945; Knotek and Schmidt, 1964). The current drinking-water standard for nitrate-nitrogen of 10 ppm was set primarily to prevent the occurrence of methemoglobinemia in infants. However, cases of infant methemoglobinemia, some resulting in death, continue to occur in rural areas (Grant, 1981; Johnson et al., 1987). Other health effects reported to be associated with nitrate contamination of ground water have included hypertension (Morton, 1971; Malberg et al., 1978), clinical methemoglobinemia in older schoolchildren (Petukhov and Ivanov, 1970), increased infant mortality (Super et al., 1981), and central nervous system birth defects (Dorsch et al., 1984), although none of these reports have been confirmed by others (Gelperin et al., 1975; Arbuckle et al., 1988). In a recent ecologic study in Nebraska, using data from the Nebraska Department of Health Registry and comprehensive ground-water data for 1984-1988, we found no trend toward higher rates of fetal death, infant death, central nervous system birth defects, or all birth defects with increasing levels of ground-water contamination by nitrates (Weisenburger et al., unpublished). Clearly, the above adverse health effects linked to ground-water contamination by nitrates, other than infant methemoglobinemia, have not been proven. However, the paucity of published research on this subject leads the author to conclude that additional studies are needed to assure the public that adverse health effects due to nitrates are unlikely.

In adults, cancers of the stomach, esophagus, nasopharynx, and urinary bladder have been associated with dietary exposure to N-nitroso compounds (Forman, 1987; Preussmann and Tricker, 1988; Forman, 1989), and children exposed to N-nitroso compounds are reported to have an increased risk of brain tumors (Preston-Martin et al., 1982). Although nitrate per se does not appear to present a cancer risk, it acts as a precursor to nitrite which forms via bacterial reduction in the saliva, atrophic stomach, or infected urinary bladder. Nitrite then reacts with nitrosatable substrates, either directly or indirectly (again via bacteria), to produce N-nitroso compounds which are potent carcinogens in experimental animals (Preussmann and Tricker, 1988; Forman, 1989). Persons drinking water with 10 ppm nitrate-nitrogen have a two-fold increased daily intake of nitrate when compared to those drinking nitrate-free water, and at 20 ppm the daily intake of nitrate increases approximately three-fold (Moller et al., 1989a). At higher levels of contamination, drinking water contributes over 80% of the dietary nitrate intake (Chilvers et al., 1984; Moller et al., 1989a). In a recent study of a Danish rural population, Moller et al. (1989b) showed that increased endogenous nitrosation is strongly associated with increased nitrate intake in drinking water and diet. These findings suggest that increased nitrate intake in drinking water may increase the risk of cancer, although epidemiologic studies of this subject are controversial (Forman, 1987; Preussmann and Tricker, 1988; Forman, 1989).

In our ecologic correlational study, we found a significantly increased incidence of non-Hodgkin's lymphoma (NHL) in Nebraska counties with 20% or more of the wells contaminated by nitrates in excess of 10 ppm nitrate-nitrogen. This finding is further supported by a similar correlation with fertilizer usage by county, whereas pesticide usage and corn production did not correlate with the occurrence of NHL. In a recent study in Iowa

(Isacson, 1988), a 60% increased risk of NHL was found in small communities (less than 1000 persons) with nitrate contamination of drinking water in excess of 5 ppm nitrate-nitrogen. Increased nitrate excretion in the urine has also been shown to have a significant positive correlation with the risk for leukemia (Chen et al., 1989). In experimental animals, N-nitroso compounds are known to be potent inducers of NHL and leukemia (Mirvish et al., 1987). These findings suggest that NHL, and possibly leukemia, should be added to the list of cancers which may possibly be induced by N-nitroso compounds.

Since small amounts of pesticides, most commonly atrazine, are often found in midwestern ground water contaminated by nitrates, the possibility that N-nitroso pesticides contribute to an increased cancer risk should be considered (Eisenbrand et al., 1975). In the laboratory, we have synthesized N-nitrosoatrazine and found it to be readily formed from atrazine and nitrite at acid pH (Weisenburger et al., 1987). N-nitrosoatrazine is a mutagen in the Ames test and a strong mutagen in the Chinese hamster V-79 assay, producing revertants 3.4 times the dimethylnitrosamine control (Weisenburger et al., 1988). Atrazine was not found to be mutagenic in either system (Weisenburger et al., 1988), although it has been reported to induce breast tumors in laboratory animals (Ciba-Geigy Corporation, personal communication). Preliminary tests in our laboratory failed to demonstrate that either atrazine or N-nitrosoatrazine are carcinogenic (Weisenburger et al., 1990), but the tests were suboptimal and further studies are necessary.

Alternatively, elevated nitrate levels in ground water may just be a geographic marker of other agricultural activities which are causal for NHL. For example, intense exposure of farmers to certain pesticides has recently been linked to an increased risk of NHL (Hoar et al., 1986; Hoar Zahm et al., 1990). Clearly, correlational studies such as this one do not prove causality. However, such studies are useful for generating hypotheses that require further investigation. Confirmation of our findings by others will be necessary before any conclusions can be drawn. Epidemiologic case-control studies evaluating nitrate intake in diet and drinking water, individual nitrosation potential, and other ground-water contaminants and potential confounding variables are clearly needed. However, it would be prudent to minimize the intake of nitrates at the present time to prevent any potential long-term health effects. Currently, there is insufficient evidence to permit raising the drinking-water standard above 10 ppm nitrate-nitrogen, whereas there are some indications that the standard provides the necessary safety factor to prevent most acute and chronic health effects of ground-water contamination. Any decision to change the standard must await the results of further research.

References

Arbuckle TE, Sherman GJ, Corey PN, Walters D, Lo B (1988) Water nitrates and CNS birth defects: A population-based case-control study. Arch Environ Health 43:162-167

Chen J, Campbell TC, Li J, Peto R (1989) Diet, Lifestyle and Mortality in China: A Study of the Characteristics of 65 Chinese Counties. Oxford University Press, Oxford

Chilvers C, Inskip H, Caygill C, Bartholomew B, Fraser P, Hill M (1984) A survey of dietary nitrate in well-water users. Int J Epidemiol 13:324-331

Comley H (1945) Cyanosis in infants caused by nitrates in well water. J Am Med Assoc 129:112-116

Dorsch MM, Scragg RKR, McMichael AJ, Baghurst PA, Dyer KF (1984) Congenital malformations and maternal drinking water supply in rural South Australia: A case-control study. Am J Epidemiol 119:473-486

Eisenbrand G, Ungerer O, Preussmann R (1975) Formation of N-nitroso compounds from agricultural chemicals and nitrite. IARC Scientific Publications 9:71-74

Exner ME, Spalding RF (1990) Occurrence of pesticides and nitrate in Nebraska's ground water. Institute of Agriculture and Natural Resources, University of Nebraska, Lincoln

Forman D (1987) Dietary exposure to N-nitroso compounds and the risk of human cancer. Cancer Surveys 6:719-738

Forman D (1989) Are nitrates a significant risk factor in human cancer? Cancer Surveys 8:443-458

Gelperin A, Moses VK, Bridger C (1975) Relationship of high nitrate community water supply to infant and fetal mortality. Illinois Med J 147:155-157

Grant RS (1981) Well water nitrate poisoning review: A survey in Nebraska 1973 to 1978. Nebraska Med J 66:197-200

Halberg GR (1988) When agrichemicals and groundwater meet: Understanding the connection. J Freshwater 11:9-11

Harrington DS, Yuling Y, Weisenburger DD, Armitage JO, Pierson J, Bast M, Purtilo DT (1987) Malignant lymphoma in Nebraska and Guangzhou, China: A comparative study. Hum Pathol 18:924-928

Hoar SK, Blair A, Holmes FF, Boysen CD, Robel RJ, Hoover R, Fraumeni JF (1986) Agricultural herbicide use and risk of lymphoma and soft-tissue sarcoma. J Am Med Assoc 256:1141-1147

Hoar Zahm S, Weisenburger DD, Babbit PA, Saal RC, Vaught J, Cantor KP, Blair A (1990) A case-control study of non-Hodgkin's lymphoma and the herbicide 2, 4-dichloro-phenoxyacetic acid (2,4-D) in eastern Nebraska. Epidemiol 1:349-356

Isacson P (1988) Proceedings of Technical Workgroup, Agricultural Occupational and Environmental Health: Policy Strategies for the Future, 18-21 Sept 1988, Iowa City

Johnson CJ, Bonrud PA, Dosch TL, Kilness AW, Senger KA, Busch DC, Meyer MR (1987) Fatal outcome of methemoglobinemia in an infant. J Am Med Assoc 257:2796-2797

Knotek Z, Schmidt P (1964) Pathogenesis, incidence, and possibilities of preventing alimentary nitrate methemoglobinemia in infants. Pediat 34:78-83

Malberg JW, Savage EP, Osteryoung J (1978) Nitrates in drinking water and the early onset of hypertension. Environ Pollut 15:155-160

Mirvish SS, Weisenburger DD, Salmasi S, Kaplan PA (1987) Carcinogenicity of 2-hydroxy-ethylnitrosourea and 3-nitroso-oxazolidine administered in drinking water to male MRC-Wistar rats: Induction of bone, hematopoietic, intestinal and liver tumors. J Natl Cancer Inst 78:387-393

Moller H, Landt J, Jensen P, Pedersen E, Autrup H, Moller Jensen O (1989a) Nitrate exposure from drinking water and diet in a Danish rural population. Int J Epidemiol 18:206-212

Moller H, Landt J, Pedersen E, Jensen P, Autrup H, Moller Jensen O (1989b) Endogenous nitrosation in relation to nitrate exposure from drinking water and diet in a Danish rural population. Cancer Res 49:3117-3121

Morton WE (1971) Hypertension and drinking water constituents in Colorado. Am J Publ Health 61:1371-1378

National Cancer Institute Annual Cancer Statistics Review including Cancer Trends: 1950-1985 (1988) National Institutes of Health Publication No. 88-2789, US Department of Health and Human Services

Nebraska Agricultural Statistics Annual Report 1981-1982 (1983) Nebraska Crop and Livestock Reporting Service, Nebraska Department of Agriculture

Petukhov NI, Ivanov AV (1970) Investigation of certain psychophysiological reactions in children suffering from methemoglobinemia due to nitrates in water. Hyg Sanit 35:29-32

Power JF, Schepers JS (1989) Nitrate contamination of ground water in North America. Agric Ecosystems Environ 26:165-187

Preston-Martin S, Yu MC, Benton B, Henderson BE (1982) N-nitroso compounds and childhood brain tumors: A case-control study. Cancer Res 42:5240-5245

Preussmann R, Tricker AR (1988) Endogenous nitrosamine formation and nitrate burden in relation to gastric cancer epidemiology. In: Reed PI, Hill MJ (eds) Gastric Carcinogenesis, Excerpta Medica, Amsterdam, p 147-162

Super M, Heese H, MacKenzie D, Dempster WS, duPless J, Ferreira JJ (1981) An epidemiologic study of well-water nitrates in a group of South West African Namibian infants. Water Res 15:1265-1270

United States Census of Agriculture (1982) Nebraska County Data for Agricultural Chemicals Used. US Government Printing Office, Washington, DC

Water Resources Data for Nebraska (1978-1979) US Geological Survey Water-Data Reports NE-78-1 and NE-79-1, US Department of Interior

Weisenburger DD (1985) Lymphoid malignancies in Nebraska: A hypothesis. Nebraska Med J 70:300-305

Weisenburger DD, Joshi SS, Hickman TI, Babcook DM, Walker BA, Mirvish SS (1987) N-nitroso-atrazine (NNAT): Synthesis, chemical properties, acute toxicity, and mutagenicity. Proc AACR 28:103

Weisenburger DD, Joshi SS, Hickman TI, Walker BA, Lawson TA (1988) Mutagenesis tests of atrazine and nitrosoatrazine: Compounds of special interest to the Midwest. Proc AACR 29:421

Weisenburger DD, Hickman TI, Patil KD, Lawson TA, Mirvish SS (1990) Carcinogenesis tests of atrazine and N-nitrosoatrazine: Compounds of special interest to the Midwest. Proc AACR 31:102

catalyzing much more rapid specific rates of reaction, typically by factors of between 10 and 100 (Leach et al., 1987). The major reason for this difference in N-nitrosation rates probably relates to a differential capacity for nitric-oxide formation from nitrite in the two groups. Since nitric oxide (NO) is easily oxidized by traces of molecular oxygen, it readily forms the potent nitrosating agents N_2O_3 and N_2O_4 (Shuker, 1988). The reduction of nitrite by denitrifiers differs considerably from that of nondenitrifiers (nitrate respirers) in that the former produce gaseous oxides of nitrogen rather than ammonium. Thus, some denitrifiers may produce free NO (and hence N_2O_3 and N_2O_4) as a more normal early product of nitrite reduction. However, the role of free NO in bacterial denitrification is not completely clear, bacteria differing in the precise details of their denitrification sequence (Garber and Hollocher, 1982). In some organisms (e.g., *P. aeruginosa*), NO may remain bound to the nitrite-reductase cytochrome as a ferrous nitrosyl which would itself be expected to be a good nitrosating agent (cf., nitroprusside) (Averill and Tiedje, 1982). Preliminary experiments of our own, however, indicate that even the nitrosating agent(s) elaborated by *P. aeruginosa* behave in a very similar fashion to mixed oxides of nitrogen (N_2O_3 and N_2O_4) (Leach et al., 1990). If this is the case, the *range* of types of NNC produced by bacteria will be more restricted than those formed by the acid-catalyzed reactions of nitrite. In particular, N-nitrosamines (NA) are likely to predominate with N-nitrosamides being absent.

4 N-nitroso Compound Formation in the Stomach

As previously stated, acid-catalyzed N-nitrosation reactions will predominate in the normal stomach, but the rate and extent of reaction will be considerably limited by the relatively low concentrations of nitrite present there (predominantly of salivary origin). In the hypochlorhydric and achlorhydric stomach, could bacterially mediated N-nitrosation reactions in some instances lead to elevated genotoxic NA levels as a result of the elevated nitrite concentrations found in such juices? Model calculations suggest that this is possible, but only where sufficient numbers (5×10^7/mL) of N-nitrosation proficient bacteria (e.g., denitrifiers) are present (Leach et al., 1987). Similar considerations probably also apply to the process of bacterial NA formation in the mouth which will also ultimately contribute to gastric NA levels. However, in addition to their carriage rate in gastric juice, several other factors have to be considered in relation to the yield of NA by proficient bacteria. For example, vitamin C, although not ordinarily reactive towards nitrite at neutral pH, is a potent inhibitor of denitrifier N-nitrosation reactions (Mackerness et al., 1989). Dietary vitamin C intake may therefore reduce not only acid-mediated NNC formation in the normal stomach but also bacterially mediated N-nitrosation in the achlorhydric stomach. Further, the models used to investigate bacterial NA formation have largely involved only relatively simple *in vitro* incubations of batch-grown cells. Experiments using continuous-culture techniques, where the metabolic state of the entire population of bacterial cells can be more rigorously controlled (Pirt, 1975), have highlighted additional factors which modulate bacterial NA formation. Thus, monocultures of the denitrifier, *P. aeruginosa*, growing continuously in a rich anaerobic growth medium containing nitrate as the terminal-electron acceptor and growth-limiting substrate (concentrations in the range 2-10 mM supplied at a dilution rate of $0.1 \ h^{-1}$) (Leach et al., 1988) in all cases catalyzed high specific rates of bacterial N-nitrosation (2000-6000 nmol N-nitroso-morpholine/h/mg bacterial protein/mL reaction mixture: cf., background rates of 2nmol/h/mL in the absence of bacteria), but when the concentration of nitrate was increased to be no longer growth-limiting (now carbon limited) the N-nitrosation activity of the culture was completely abolished (Table 1). Thus, in this system designed to model particular aspects of the achlorhydric stomach (that is, a rich nutrient status, anaerobic conditions, a realistic dilution rate and a supply of nitrate), bacterially mediated N-nitrosation still proved an important feature of this bacterium, but only when the cells were adapted to conditions of nitrate-limited growth (i.e., the rate of nitrate supply in relation to that of its other growth requirements critically affected its N-nitrosation activity).

Table 1. Effect of nitrate concentration supplied to the continuous culture model on the N-nitrosation activity of *P.aeruginosa* BM1030

Nitrate concentration (mM)	Growth yield dry wt. (mg/mL)		N-nitrosation activity nmol NMOR/h/mg protein/mL
2	limited	0.050 ± 0.016	3000 ± 1333
3.5	"	0.098 ± 0.025	3750 ± 1300
5	"	0.109 ± 0.020	5833 ± 1754
7.5	"	0.140 ± 0.020	4216 ± 1927
10	"	0.235 ± 0.030	3692 ± 1000
250	excess	0.244 ± 0.015	not detected

N-nitrosation activity per unit mass of total cellular bacterial protein determined by standardized "off-line" assay procedure (Leach et al., (1987). NMOR: N-nitrosomorpholine, growth yield: mass of washed and rigorously dried bacterial cells harvested from a unit volume of culture medium from chemostat.

PANEL DISCUSSION: HEALTH CONSEQUENCES

D.D. Weisenburger, Panel Chair and Reporter
Department of Pathology and Microbiology
Eppley Institute for Research in Cancer and Allied Diseases
University of Nebraska Medical Center
Omaha, Nebraska 68198-3135 U.S.A.

Panel Members: M. Crespi,[1] S. Mirvish,[2] H. Møller,[3] J. Hotchkiss,[4] S.A. Leach,[5] and
D. Forman[6]

1 Panel Consensus on Potential Health Consequences of Elevated Nitrate Levels

The panel reached the following consensus on the potential health consequences of elevated
nitrate levels in drinking water:

1. Currently, methemoglobinemia in infants under six months of age is the only illness
which is clearly caused by the intake of drinking water with elevated nitrate levels. The
current standard of 10 ppm nitrate-nitrogen (45 ppm nitrogen) was set to prevent the occurrence of infant methemoglobinemia and provides a reasonable margin of safety to do so.

2. Many N-nitroso compounds are potent carcinogens in experimental animals. Increasing
nitrite intake in animal studies results in increased endogenous formation of carcinogenic
nitrosamines in proportion to the square of the nitrite concentration, whereas nitrosamide
formation is proportional simply to the nitrite concentration.

---

[1] Department of Environmental Carcinogenesis, Epidemiology and Prevention, National
Cancer Institute "Regina Elena," Rome, Italy.
[2] Eppley Institute for Research in Cancer and Allied Diseases, University of Nebraska
Medical Center, Omaha, Nebraska, U.S.A.
[3] Danish Cancer Registry, Institute of Cancer Epidemiology, Copenhagen, Denmark.
[4] Institute for Comparative and Environmental Toxicology, and Institute of Food Science,
Cornell University, Ithaca, New York, U.S.A.
[5] Pathology Division, Center for Applied Microbiology and Research, Public Health
Laboratory Service, Wiltshire, UK
[6] Cancer Epidemiology Unit, Imperial Cancer Research Fund, University of Oxford, Oxford,
UK

NATO ASI Series, Vol. G 30
Nitrate Contamination
Edited by I. Bogárdi and R. D. Kuzelka
© Springer-Verlag Berlin Heidelberg 1991

3. In humans, the nitrosoproline test has demonstrated that increasing levels of nitrate intake in drinking water lead to increased endogenous nitrosation with the increased formation of the noncarcinogenic nitrosamine, nitrosoproline, which is excreted in the urine. There is no apparent threshold for this reaction.

4. Persons who drink water with 10 ppm nitrate-nitrogen have twice the nitrate intake of those drinking nitrate-free water; at 20 ppm, daily nitrate intake increases approximately three-fold. At levels above 20 ppm, drinking water contributes over 80% of the dietary nitrate intake.

5. Conclusive proof of a carcinogenic effect in humans of nitrate or N-nitroso compounds is currently lacking. The results of epidemiologic studies are contradictory and inconclusive, with many of the studies being flawed and inadequate. The majority of evidence by which nitrate exposure has been associated with an increased cancer risk is derived from correlational studies which, by their nature, provide only weak evidence. Therefore, such evidence should not be used to establish a cause and effect relationship.

6. At the present time, nitrate exposure in drinking water cannot be implicated or excluded as a causative factor for certain types of cancer. Currently, risk estimates for given nitrate intake levels cannot be reliably calculated, nor can an absolute safety standard be established. However, reduction of the current standard for nitrate-nitrogen in drinking water (10 ppm) is unlikely to significantly diminish any potential cancer risks. Therefore, no changes in the current standard should be made until more conclusive evidence of adverse health effects is forthcoming.

2 Panel Recommendations

The following recommendations were put forward:

1. Since the formation of N-nitroso compounds in humans is a complex process, additional basic research to determine the importance of the various biologic variables is needed (i.e., available dietary substrates for nitrosation, gastric pH, gastric volume and emptying time, vitamin C and other modifiers, bacterial flora, identification and carcinogenicity of N-nitroso compounds formed, etc.).

2. Well-designed, individual-based epidemiologic studies of the case-control or cohort types are needed. If possible, such studies should include information about total nitrate and ascorbate intakes, and should incorporate the use of biologic markers of exposure (i.e., 24-hr nitrate excretion, nitrosoproline test, carcinogenic N-nitroso compounds, DNA adducts in urine, etc.).

3. Specific population subgroups that are potentially at increased risk for the development of cancers associated with N-nitroso compounds should be identified and studied (i.e., those with chronic gastritis, achlorhydria, partial gastric resection, chronic esophagitis or chronic urinary tract infection, those taking drugs that decrease gastric acidity, smokers, etc).

4. Research should also be directed at potential health effects other than cancer which have been identified in the medical literature (i.e., increased infant mortality, birth defects, methemoglobinemia in older children, hypertension, etc.).

5. Significant funds should be allocated to support the research needed to answer the many questions about the potential health consequences of excessive nitrate intake.

6. Because of many uncertainties, it remains prudent to limit the amount of nitrate in drinking water to the current standard. The agricultural community should be encouraged to develop innovative strategies to prevent contamination of ground water by nitrate in the future.

IV. CONTAMINATION CONTROL

Although most of the measures outlined above will be helpful in limiting nitrate leaching, some have serious economic and/or other disadvantages. Good agricultural practices on their own will not be sufficient to reduce nitrate concentrations in ground water where they are already above the EC limit. Estimates from several sources of the effect of reductions in nitrogen applications on leaching losses are shown in Fig. 2 (HoL, 1989). Reductions of 20%, for example, are estimated to reduce leaching by amounts ranging from 0 to 40%. In the drier parts of eastern England, leaching rates above about 20 kg N/ha will produce nitrate concentrations above 11.3 mg NO_3-N/L in recharge to ground water. Thus, reductions in the

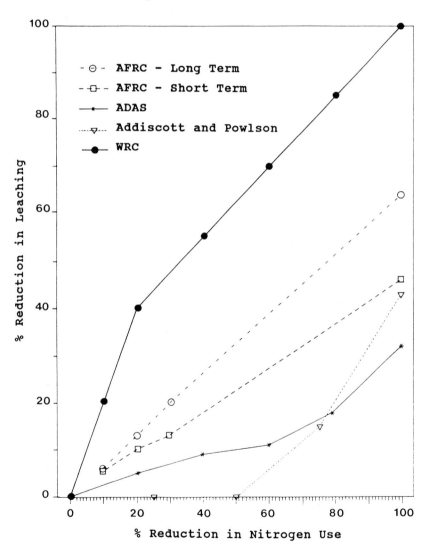

Fig. 2. Effect of nitrogen fertilizer reduction on leaching from winter wheat

leaching rate of some 50% are required to achieve acceptable concentrations. Most estimates suggest this cannot be achieved without an almost total elimination of fertilizer usage.

Given these circumstances, the only approach remaining is one of radical changes in agricultural activities in the designated vulnerable zones around public-supply sources, introducing land uses from which nitrate leaching losses are small. The options are limited to grassland or afforestation, both of which are currently the subject of studies by the British Geological Survey.

2.2.3 Land-use Changes

Most of the investigations of nitrate leaching from permeable soils to ground water in the 1970s and early 1980s were concentrated on arable land. The few investigation boreholes that were drilled on grassland sites indicated minimal leaching of nitrate from unfertilized or lightly-fertilized grassland (Fig. 3). Grassland production responds more to higher rates of

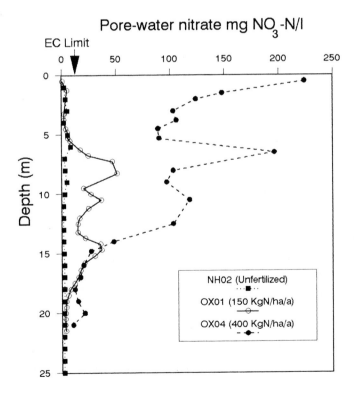

Fig. 3. Unsaturated zone pore-water nitrate profiles for selected grazed grassland sites

nitrogen fertilizer than do other crops. Optimum application rates are in the range of 350 to 450 kg N/ha/yr for both herbage and animal production, and application rates have increased considerably in the last 30 years. The issue of nitrate leaching from grazed grassland was highlighted by Ryden et al. (1984), who estimated an equivalent leaching loss of 160 kg N/ha/yr from an application of 420 kg N/ha/yr, significantly greater than that beneath a comparable cut grass sward, from which most of the nitrogen is removed in the hay or silage.

During the period 1987 to 1989, studies of nitrate leaching to ground water have been carried out by the British Geological Survey at four grassland sites on the outcrop of the Chalk aquifer in southern England, with emphasis on intensively managed grazed grassland. Nitrate concentrations in the unsaturated zone of the aquifer were established by means of pore-water sampling from undisturbed cores retrieved from shallow investigation boreholes.

The results show (Fig. 3) that nitrate leaching from grazed grassland is significant and almost always results in nitrate concentrations in excess of 10 mg NO_3-N/L in ground-water recharge. Where the grassland is intensively managed, with applications of nitrogen fertilizer of 400 kg N/ha/yr or more, ground-water nitrate concentrations often exceed 50 mg NO_3-N/L (Fig. 3). These concentrations are greater than those observed beneath intensively cultivated arable land with comparable soils and similar climate. The data imply average leaching losses of 75 to 150 kg N/ha/yr from intensively managed grazed grassland, which represents a loss equivalent to 20 to 40% of the nitrogen inputs. Leaching of nitrate from grazed grassland receiving more than about 100 kg N/ha/yr is likely to produce nitrate concentrations above the EC limit in ground-water recharge. Grassland is, therefore, only appropriate as a land-use option for controlling nitrate leaching to ground water if it is lightly fertilized and/or harvested rather than grazed.

The establishment of deciduous woodland would seem to be an attractive measure for areas designated as vulnerable to nitrate pollution. There is, however, little reliable information available on the likely impact that this could have on the leaching of nitrate to ground water. Most studies in Britain of the effects of afforestation have been concerned with the impact of coniferous plantations on the water balance in upland areas, and only more recently have the impacts on water quality been studied. Results of catchment water balances show that typically an extra 20% of the rainfall is lost by evaporation from an upland forested catchment, mainly by an increase in the interception of rainfall; i.e., by evaporation directly from the surface of the wet leaves rather than transpiration through the stomata. In the drier low-land areas in the east and south of the country, the impact on ground-water resources of extra evaporative losses of this magnitude need to be considered.

Very little is known about the impact of lowland afforestation on the quality of ground water, particularly in the early stages after converting from arable land. The immediate benefit would be the reduction in fertilizer use, but it is not clear how efficiently the young trees would use up the excess nitrogen already present in the soil. A current British Geological Survey study involves sampling of interstitial water from undisturbed drill cores taken from the unsaturated zone beneath established woodland. This has shown that nitrate concentrations are extremely variable in all three dimensions, ranging from less than 1 mg NO_3-N/L to more than 25 mg NO_3-N/L. The smallest nitrate concentrations have been found beneath large trees at the edge of woodland and have been associated with relatively large concentrations of chloride (more than 100 mg/L). These are situations where evaporation is unusually high due to the exposed edge position and where nitrate, but not chloride, is readily assimilated by the growing trees. Some profiles are characterized by smooth, broad peaks of nitrate concentration extending over several meters depth. High nitrate concentrations are frequently found in areas of old woodland where the tree density is low due to natural die back. It appears that during the most active growing period, the combined inputs of nitrogen from atmospheric deposition and mineralization of soil organic matter are efficiently cycled and largely remain in the soil and plant biomass. However, when the nutrient cycle is broken, the nitrogen which has accumulated is readily leached from the soil and is a potential source of increased nitrate concentration in ground-water recharge. Providing woodland is sensibly managed, it could be a suitable control measure in vulnerable zones. However, to make the existing EC legislation for agricultural Set-Aside effective in relation to nitrate leaching, the scheme would have to be made permanent and more easily targeted on vulnerable areas.

2.3 Response Times for Agricultural Control Measures

In areas where ground-water nitrate concentrations are already high, even radical land-use changes cannot be expected to have immediate beneficial effects on ground-water quality. There is a natural time-lag in the ground-water system such that past agricultural practice will continue to affect nitrate concentrations in pumped ground water for many years. A recent desk study (DoE, 1988) of ten catchments using a mathematical model developed by Oates (1982) illustrated the importance of the time lag in different aquifers. For three Chalk catchments in which the nitrate levels are already above 11.3 mg NO_3-N/L, a reduction to this level could not be achieved in 50 years even if the entire area of the catchments were immediately converted to unfertilized grass. In contrast, the limestone catchments which were examined responded much more quickly.

The reasons for the time lag become clear from a consideration of the properties of the three main British aquifers (Table 1). Taking typical porosity values, and for the range of effective

Table 1. Characteristics of principal British aquifers

	Cretaceous Chalk	Triassic Sandstone	Jurassic Limestone
Lithology	microporous carbonate	quartz grains with sparse silica or carbonate cement	microporous or dense carbonate
Ground-water flow regime	fissure and matrix	intergranular with some fissure	fissure
Matrix porosity (%)	25-45	15-35	10-25
Characteristic pore size (μm)	0.2-1	5-50	0.05-0.5
Matrix hydraulic conductivity (m/d)	2×10^{-4} to 5×10^{-3}	5×10^{-1} to 10	up to 5×10^{-4}
Effective rainfall (mm)	150-350	200-350	150-250
Unsaturated zone flow rates (m/yr)	0.3-1.4	0.6-2.3	0.6-2.5
Unsaturated zone thickness (m)	5-50	10-40	5-25
Unsaturated zone residence times (yr)	4-120	3-65	2-40
Fissure spacing (m)	5-10 (macro) 0.1 (micro)	Very variable and difficult to generalize	1-5 (macro)
Natural ground-water flow velocity (m/d)	1-10		5-25
Aquifer thickness (m)	up to 450	up to 500	20-35

rainfall over the aquifer outcrop, unsaturated zone flow rates can be estimated for each aquifer (Table 1), assuming most of the flow is through the rock matrix. Sequential profiling of tritium (Geake and Foster, 1989) suggests vertical movement in the unsaturated zone of the Chalk of 0.5-1.5 m per year, confirming the estimates using conventional meteorological models (Grindley, 1969) and the aquifer properties given in Table 1. In areas of the Chalk where fissure flow is dominant and in the Lincolnshire Limestone, the tritium peak is not well preserved and the method cannot be readily applied. In these situations, the proportion of so-called "fast" or "by-pass" flow is greater and the unsaturated zone residence times are nearer the lower ends of the ranges given in Table 1.

Saturated zone flow rates and the volume of storage in the aquifer are also important considerations in determining the response time at ground-water sources. In this respect, the Lincolnshire Limestone is a relatively thin aquifer (Table 1) with modest volumes of ground-water storage, whereas the Chalk and Triassic Sandstones are much thicker (Table 1). Saturated zone storage thus has a much greater buffering effect on response times in the Chalk and Sandstones than in the Limestones. The net effect of unsaturated and saturated zone residence times is that the longest response times can be expected in the Chalk, and the shortest in the limestones, with the sandstones falling between the two.

2.4 Implementation of Land-use Changes

Protection of ground water by the introduction of land-use changes in vulnerable zones could comprise limited modifications to agricultural practice over the whole aquifer outcrop or more radical changes in a limited protection zone around public supply sources. Taking the most stringent interpretation of the EC Directive, in which not even occasional peak nitrate concentrations above 11.3 mg NO_3-N/L are permitted, it has been estimated that much of southern and eastern England and parts of the midlands would have to be declared vulnerable (HoL, 1989; see also Fig. 1). Similarly, the Water Authorities Association estimated that on this interpretation, four million hectares would have to be designated in respect of surface waters and, on any interpretation of the Directive, they estimated the vulnerable zones in respect of ground water would cover 300,000 ha (HoL, 1989).

The two approaches have different financial, practical and social implications. Limited

Conversion to grassland is only suitable in this respect where it is unfertilized or lightly fertilized. Intensively managed grassland with high nitrogen applications and heavy grazing produces nitrate leaching equal to or greater than cereal cultivation. Preliminary indications are that farm afforestation could be effective in limiting nitrate leaching, but the increased evaporation has resource implications which need to be taken into account.

Even where major land-use changes are implemented in the catchments of ground-water sources, nitrate concentrations in ground water will reflect past agricultural practices for many years. In general, the longest response times will occur in the Chalk and the shortest in the Lincolnshire Limestone, with the Sandstones intermediate between these.

The costs, both directly to water undertakings and their consumers and indirectly to the national economy, of compliance with the EC Directive are considerable. Because of the long time lags before preventive measures become effective, curative measures must be implemented in the meantime to bring high concentrations below the EC limit and to keep rising nitrate concentrations below this limit. Local hydrogeological conditions, climate, farming systems and existing water supply configurations greatly affect the comparative costs of preventive and curative measures. The viability of each will vary greatly from place to place, and varying combinations of the two are likely to be employed.

Acknowledgements

This paper is published by permission of the Director of the British Geological Survey (NERC). The studies by the British Geological Survey of nitrate in ground water have been funded principally by the Department of the Environment to 1989 and by the National Rivers Authority from 1990. Reference is made to the results of interdisciplinary national study groups in which the British Geological Survey participated. The authors gratefully acknowledge the contributions and valuable discussions with their colleagues Dr D.G. Kinniburgh and Ms. J.M. Parker.

References

Croll BT, Hayes CR (1988) Nitrate and water supplies in the United Kingdom. Environmental Pollution 50:163-187
Department of the Environment (1986) Nitrate in water, Pollution Paper No 26, Her

Majesty's Stationery Office, London

Department of the Environment (1988) The nitrate issue, Her Majesty's Stationery Office, London

Foster SSD, Bridge LR, Geake AK, Lawrence AR, Parker JM (1986) The groundwater nitrate problem, Hydrogeological Report 86/2, British Geological Survey, Wallingford

Geake AK, Foster SSD (1989) Sequential isotope and solute profiling in the unsaturated zone of British Chalk. Hydrological Sciences Journal 34:79-95

Grindley J (1969) The calculation of actual evaporation and soil moisture deficits over specified catchment areas. Hydrological Memo 38, Meteorological Office, London

House of Lords Select Committee on the European Communities (1989) Nitrate in water, Session 1988-89, 16th Report, HL Paper 73, Her Majesty's Stationery Office, London

Oakes DB (1982) Nitrate pollution of groundwater resources:mechanisms and modelling. In K-H Zwirnmann (ed) Non-point Nitrate Pollution of Municipal Water Supply Sources:Issues of Analysis and Control. The International Institute for Applied Systems Analysis, Laxemburg, Austria, p 207-230

Ryden JC, Ball PR, Garwood AE (1984) Nitrate leaching from grassland. Nature 311:50-54

Severn-Trent Water (1988) The Hatton catchment nitrate study, Severn-Trent Water, Birmingham nitrates are converted into ammonia which, in turn, must be removed using ion exchange or similar methods) (Sova, 1986; Dahab, 1987).

2 Nitrate Removal Using the Ion-Exchange Process

2.1 The Process

The ion-exchange process involves the exchange of ions in solution (i.e., contaminated water) with chemically equivalent numbers of ions associated with the exchange material (i.e., the resin). Ion-exchange materials include naturally occurring substances such as clays and synthetic substances including a variety of chemically prepared resins. Currently used materials in the water-treatment field are in the form of porous granules or beads. The structure of these resins usually consists of an interconnected network of hydrocarbons to which are attached soluble ionic functional groups. The exchangeable ions are either negatively charged (anions) or positively charged (cations) depending on the nature of the resin. The number of exchangeable ions per unit volume or mass determines the exchange capacity of the material (Weber, 1972).

The mechanism involved in the removal of nitrate ions from water is the replacement of these ions with chloride ions when water containing nitrates is passed over the resins. The process utilizes either strongly basic or weakly basic anion exchangers. Anion exchangers or resins containing functional groups made up of weak amine bases, derivatives of ammonia, are called weakly basic. Those derived from quaternary ammonia compounds are referred to as strongly basic (Andersen, 1980, 1981).

2.2 Limitations

Ion exchange offers great potential for application at small and medium-sized treatment plants (Dahab, 1987; Lauch and Guter, 1986; Montgomery, 1985). Application would be very similar to domestic (or commercial) ion-exchange softeners currently available in the market place. However, two basic problems must be fully addressed before widespread application can be realized.

The first problem is to provide a resin of high selectivity for nitrates over other ions that are commonly present in most ground-water supplies, often at higher concentrations than nitrates.

Exchange resins exhibit a degree of preference or selectivity for various ions depending upon the concentration of ions in solution. Normally, ions with higher valences (electrical charges), higher atomic weights and smaller radii are preferentially removed by exchange resins from solution. For anions the normal order of preference is (Weber, 1972):

$$PO_4^{3-} > SO_4^{2-} > HPO_4^{2-} > NO_3^- > HCO_3^- \qquad [1]$$

One of the major problems associated with the use of the ion-exchange process for removing nitrates from water is that although the radius of the nitrate ion is slightly less than that of sulphate, most resins have definite preference for the sulphate ion because of its higher valence. This preference limits the effectiveness of ion exchange whenever the sulfate content of the water supply exceeds that of the nitrates (Andersen, 1980). Some researchers have reported the development of a series of nitrate-selective resins (Guter, 1982, 1987; Lauch and Guter, 1986; Van Der Hoek and Klapwijk, 1987, 1988), but the stability and economy of these resins have not been confirmed in prolonged use. Nitrate-removal experiments at a California site (Lauch and Guter, 1986) indicated that a treatment cost (for nitrate removal only) of at least $0.18 per one thousand gallons of product water is to be expected (1985 dollars) when treating well waters with a nitrate-nitrogen concentration of less than 20 mg/L.

The second problem is to provide an adequate resin regenerant so that regenerant disposal does not itself become a problem. Currently, regenerant disposal may account for a major fraction of the overall cost of the process (Weber, 1972; Clifford, 1987). Alternatives available for regenerant disposal include discharge to the local municipal waste treatment works, application to land, and transport to other treatment works. In a recent California installation, Guter (1990) reported that a fixed-bed biodenitrification system is used to treat the brine generated by the ion-exchange process.

The molar replacement of a large part of the nitrate and sulphate ions, and to a lesser extent of the bicarbonate ions, can increase the corrosiveness of the treated water because of the increase of chloride ions and their possible modification of the calcium-carbonate balance. If appropriate measures are not taken to counteract this problem, it can cause corrosion problems in the downstream distribution system.

Other disadvantages of the ion-exchange process include the addition of undesirable treating part of the influent water and then blending it with raw water so that the nitrate concentration from each plant meets the drinking water standard (Guter, 1990; Guter and Kartinen, 1989).

Each plant consists of three reaction vessels designed to operate in staggered fashion so that two vessels are in the service mode and the third is in the regeneration mode. The plants are microprocessor-controlled and reported to be fully automatic. The anion-exchange resin used in both plants substitutes chloride ions for nitrate ions and removes sulphate ions in preference to all. No especially prepared resins or any special mechanisms for recycling or regeneration to minimize brine are used. Brine is disposed of at the municipal wastewater treatment plant and then discharged to agricultural land that is used to grow animal feed and cotton. However, problems resulting from the disposal of the brine, which includes nitrates removed from the treated water coupled with salt, are greatly feared (Guter and Kartinen, 1989). Therefore, the need to develop methods of reuse and recovery of wastewater and salt as well as the reduction of discharge volume is apparent.

Ion-exchange plants have recently been built in the cities of Crescenta and Grover City, California (Guter, 1990). No operational data is available from these two locations yet.

A novel technique combining ion-exchange and biological denitrification was described by Van Der Hoek et al. (1988) and Van Der Hoek and Klapwijk (1988). This combination was shown to be effective for nitrate removal in pilot-scale studies. In this process, nitrate is removed from ground water by ion-exchange, and biological denitrification is used in conjunction with regeneration of the nitrate-loaded resin. The regeneration process can be carried out using sodium chloride (NaCl) or sodium bicarbonate (NaHCO$_3$) solutions. For example, when using the bicarbonate solution as the regenerant, the solution first passes through the ion-exchange column where nitrate ions are exchanged for bicarbonate ions, and then is forced through the denitrification reactor where denitrifying bacteria convert nitrates into nitrogen gas. The regenerant is recirculated through the ion-exchange column and the denitrification reactor until the ion-exchange resin is sufficiently loaded. The regeneration thus takes place in a closed system. The reported added advantages of this system over ordinary ion-exchange systems include the reduction of the otherwise voluminous brine, no direct contact of biological process with the ground water, and no risk of nitrite formation.

Another method was initially developed between 1978 and 1982 at the Karlsruhe Nuclear Research Center in West Germany (Hagen et al., 1986) and then further developed for commercial application by a private business concern. This process consisted of using a combined weakly acidic exchanger in the free-acid form with an anion exchanger in the HCO_3^- form in a mixed bed. The exchangers remove neutral salts from solution, including nitrates, and liberate equivalent amounts of carbon dioxide when in contact with raw water. The exhausted resins are regenerated by bringing them in contact with concentrated carbon dioxide solution. One of the reported advantages of this process is that the resins do not need to be regenerated separately as in other mixed-bed processes because of the favorable influence each has on the other. Other reported advantages of this process include the required use of volumetric ratio of two exchangers according to the type of demineralization: The resin make-up can be modified easily to suit the raw water salt concentration (Hagen et al., 1986).

3 Nitrate Removal Using Reverse Osmosis

3.1 The Process

Reverse osmosis is a process whereby ionic species (e.g., nitrates) present in the water supply are removed by forcing the water across a semipermeable membrane. In this process, the water supply in the reverse-osmosis cell is subjected to pressures exceeding its corresponding osmotic pressure. Such pressures can easily reach 300 to 400 psig when treating brackish water and up to 1000 psig when reverse osmosis is used in desalinating seawater (Montgomery, 1985; Weber, 1972).

Membranes used in the manufacture of reverse-osmosis units are often made of cellulose acetate and similar polymeric materials and must be made to withstand high pressures. These membranes generally do not exhibit high selectivity for any given ion although the degree of salt rejection seems to be directly related to the valency of ions present in the water supply. In consequence, the reverse-osmosis process generally results in removal of many ionic species (i.e., minerals) that are present in the water supply including nitrates (Applegate, 1984, Montgomery, 1985). This characteristic means that not only a portion of the mineral content of the water supply is removed but, more importantly, it means that only a fraction of the nitrate present in the water should actually be expected to be removed.

Common problems associated with reverse-osmosis membranes include fouling, compaction, hydrolytic deterioration, and concentration polarization (Eisenberg and Middlebrooks, 1984). These problems usually are caused by deposition of soluble materials, organics, suspended solids, colloidal particles and other contaminants on the membrane surface, excessive temperature, pH variation outside tolerance levels, biological and chemical attacks, and chlorine exposure. However, these problems can be reduced by pretreatment (e.g., filtration, pH adjustment, carbon adsorption, chlorination, coagulation of colloidal matter or a combination of these), adjustment of fluid turbulence and module conversion, etc.

Despite the high energy input required to produce the pressures needed to drive reverse-osmosis units, the process can be fairly comparable to other processes under some conditions, and therefore merits a close investigation when considering potential processes for nitrate and other dissolved-solids removal (Baier et al., 1987). There are some full-scale reverse-osmosis plants across the United States, most of which were built for total dissolved solids (TDS) reduction (Ashton, 1986; Dykes and Conlon, 1989; AWWA, 1989; Buros, 1989). It should be noted that it might be possible to economically justify reverse osmosis as a very viable nitrate-removal or reduction process by allocating a significant portion of the process capital and operating costs to the reduction of other dissolved solids that are likely to be present in the ground-water supply. This type of cost-allocation system is plausible particularly if the water supply requires treatment for hardness (i.e., softening) or reduction of other chemical constituents such as sulfates and chlorides.

The performance of a reverse-osmosis system is often measured in terms of flux, conversion, rejection, and salt flow rates. With respect to nitrates, an economically justifiable system would necessitate high nitrate rejection by the membranes, high water flux at low driving force (low transmembrane pressure difference), a high recovery rate, and low salt leakage.

3.2 Brief Discussion of Some Reverse-Osmosis Systems

Particular attention has been paid in France to the problem of nitrate contamination. In the beverage industry, ion-exchange systems were initially emphasized; but after several pilot-system tests, the emphasis shifted to advanced-membrane processes (Marquardt, 1987). In the following discussion, some reported experience with such treatment systems is presented.

In one reverse-osmosis treatment plant, well water characterized by very high magnesium hardness and a nitrate content of 21.2 mg/L is treated. The plant has been in operation for more than five years and operates 135 to 150 hours per week. Beer of good quality was reported to be produced from the product water (Marquardt, 1987).

In another case, well water, also high in total dissolved solids (TDS) and a nitrate content of 24.1 mg/L was reported to be treated for the production of good quality beer. This reverse-osmosis system was reported to have been producing water for more than six years, and operating 120 to 140 hours per week. In this system, the raw water is pretreated using flocculants, filtered, and then pumped through reverse-osmosis cells (Marquardt, 1987).

A pilot plant for the separation of nitrates by membrane processes was reported by Rautenbach et al. (1986). This plant was reportedly operated at almost "zero discharge" condition, had a capacity of 2 m^3/h, employed a spiral-wound modules with composite membranes and operated at a feed pressure of about 14 bar. Pretreatment consisted of 5 μm cartridge filters and acid dosing in order to prevent carbonate scaling. Recovery rates were up to 75-80% and module cleaning with diluted citric acid was scheduled on a 2 to 3 month basis. The observed nitrate rejection rate was about 93-95% (Rautenbach et al., 1986).

4 Nitrate Removal Using Biological Denitrification

4.1 The Process

Biological Denitrification is a well established process in the realm of wastewater treatment. However, this process has not been introduced to the field of water treatment in any significant scale. There are several pilot and full-scale demonstration plants being operated in Europe (Guantlet, 1981; Kreevoy and Nitsche, 1982; Philipot, 1982, 1985, 1988; Philipot and De Larminat, 1988; Sharefkin et al., 1985; Gayle, et al., 1989). In addition, there is one small demonstration plant currently being evaluated in the United States. The primary reason behind the slow transfer of technology from the wastewater treatment to potable water treatment is the obvious concern over potential bacterial contamination of the treated water supply. This is a legitimate concern and must be kept in mind when designing such treatment processes for water treatment.

Biological denitrification has been suggested as a reasonable process for nitrate removal from contaminated water. Dahab (1985, 1987) and Dahab and Lee (1988) reported on the potential for using biological denitrification for nitrate reduction in ground-water supplies in laboratory-scale experiments at the University of Nebraska-Lincoln. The results indicated that this process can be expected to reduce the nitrate concentration in the influent water supply from as high as 100 mg/L (as N) to levels within the 1.0 mg/L (as N) range. These removals translate into an efficiency rate of nearly 100%, which is not matched by other processes available for nitrate reduction. However, some residual soluble as well as insoluble organic matter should be expected in the denitrified water supply. Further treatment should reduce these solids to levels sufficient to meet drinking water standards.

In biological denitrification, facultative microorganisms are contacted with the water supply containing nitrates and an added carbon source in an anoxic environment. Under these conditions, the bacteria utilize nitrates as a terminal electron acceptor in lieu of molecular oxygen. In the process, nitrates are reduced to harmless nitrogen gas. An extraneous carbon source is necessary since it supplies the energy required by the microorganisms for respiration and synthesis. Most denitrification studies have used methanol (CH_3OH) as the carbon source. If a simple carbon source such as methyl alcohol or acetic acid is chosen, then the biological solids produced during the process will be correspondingly low; this is a useful characteristic since the overall sludge production is minimized. The stoichiometric relationships describing this process are written as follows (Metcalf and Eddy, 1979):

Bacterial energy reaction, step 1:

$$6 \ NO_3^- + 2 \ CH_3OH \rightarrow 6 \ NO_2^- + 2 \ CO_2 + 4 \ H_2O \qquad [2]$$

Bacterial energy reaction, step 2:

$$6 \ NO_2^- + 3 \ CH_3OH \rightarrow 3 \ N_2 + 3 \ CO_2 + 3 \ H_2O + 6 \ OH^- \qquad [3]$$

Overall energy reaction:

$$6 \ NO_3^- + 5 \ CH_3OH \rightarrow 3 \ N_2 + 5 \ CO_2 + 7 \ H_2O + 6 \ OH^- \qquad [4]$$

Bacterial synthesis reaction:

$$3 \ NO_3^- + 14 \ CH_3OH + CO_2 + 3H^+ \rightarrow 3 \ C_5H_7O_2N + H_2O \qquad [5]$$

In the bacterial synthesis reaction, nitrate is converted to cell tissue. The formula $C_5H_7O_2N$ is a representation of the cell tissue formed. In practice, 25 to 30% of the methanol required for this reaction is used for bacterial synthesis (Metcalf and Eddy, 1979). On the basis of experimental laboratory studies, the following empirical equation was developed to describe the overall nitrate-removal reaction:

$$NO_3^- + 1.08\ CH_3OH + H^+ \rightarrow 0.065\ C_5H_7O_2N + 0.47\ N_2 + 0.76\ CO_2 + 2.44\ H_2O \qquad [6]$$

If only nitrate is present, as in most nitrate-contaminated natural water supplies, Equation [6] can be used to determine the overall methanol requirement. However, if some nitrite and dissolved oxygen are present, the methanol requirement is correspondingly higher.

Denitrifying bacteria require a carbon source for respiration and growth, and a wide variety of organic compounds have been used including methanol, ethanol, and glucose. While the types of organic compounds may affect the biomass yield, the choice is generally based on economic comparison. The availability of ethyl alcohol from agricultural sources could make this carbon source a strong candidate for denitrification systems. One important factor is the presence of dissolved oxygen in the waters and its inhibiting effects. To effect denitrification, the oxygen concentration must be reduced to a level which is low enough to avoid inhibition or repression of nitrate reductase. Consequently, the amount of electron donor added must be equal to that needed to remove the oxygen as well as the nitrate and nitrite.

4.2 Advantages of Biodenitrification

Nitrate removal using biodenitrification has several advantages:

1. The process is cheaper to install with comparable operation and maintenance cost to other treatment alternatives.
2. The excess biological growth produced as waste is much easier and less expensive to dispose of than waste salts and brines from other methods.
3. The process is extremely effective in reducing nitrates to virtually near zero concentration in the treated water regardless of their concentration in the raw influent water.
4. Process stability is excellent particularly when using static-media reactor systems (i.e., biofilm systems).

5. The process does not impart excess undesirable chemicals such as chlorides to the treated water.

6. Biological treatment, in general, is probably better suited to the removal of various toxic and hazardous micro-pollutants than most physical-chemical systems (Rittmann and Huck, 1989; Bouwer and Crow, 1988)

4.3 Disadvantages and Limitations

Disadvantages of biodenitrification include the following:

1. Extreme variations in raw water characteristics, such as dissolved oxygen, total organic carbon (TOC), nutrient concentrations, pH, temperature and the presence of inhibitors can contribute to performance variability. Because ground-water supplies are generally uniform in quality and temperature, this disadvantage may not be consequential to biodenitrification.

2. Oxygen can be inhibitory to the process. The minimum oxygen concentration which inhibited or repressed nitrate reductase is not clear but a range of 0.1 to 0.22 mg/L was reported (Rittmann and Huck, 1989). Therefore, to provide effective denitrification, the oxygen concentration must be controlled in order to prevent inhibition.

3. The carbon-source level is critical in denitrification operations, since an insufficient supply may result in high levels of nitrates or nitrites, while overdosing will probably result in high concentrations of residual carbon in the treated water (Dahab and Lee, 1988). The presence of a carbon source in the treated water may necessitate additional treatment and cause high disinfectant demand (Bouwer and Crow, 1988; Philipot, 1985).

4. Products of microbial activity, such as endotoxins, soluble microbial products and incompletely degraded organic compounds may be imparted to the treated water.

5. Start up periods of biological processes are generally long and proper maintenance and monitoring of biomass accumulation and composition is required. However, excellent performance of biological treatment units can be achieved by proper application of process principles, regular laboratory testing and frequent monitoring and assessment (Bouwer and Crow, 1988).

A summary of some reported studies on biodenitrification for water treatment is provided in

Table 2 below for conventional systems using suspended or attached growth systems. A short discussion of *in situ* systems follows.

Table 2. A summary of biodenitrification nitrate-removal processes

Process Description	Location	Influent Concentration NO_3/L	% Removal	Reference
Sulphur lime-stone filtration, $Q=35$ m³/h (PS)	Monterland, Netherlands	70-75	95-100	Kruithof et al., 1988
Fixed-film reactor biological filtration $Q=400$ m³/h (FS)	Guernes, France	40-65	---	Ravarini et al., 1989
Biological down-flow filters, (FS)	Eragny, France	60	84	Philipot, 1982; Philipot et al., 1985
Upflow static-bed reactor (PS)	Lincoln, Nebraska	100	95-100	Lee, 1987; Dahab and Lee, 1988
"Nitazur" upflow. $Q=50$ m³/h (FS)	Chateau Landon, France	80-90	72	Richard and Burriat, 1988
"Nitrazur" upflow, $Q=70$ m³/h (FS)	Champfleur, France	72	65	Richard and Burriat, 1988
Slow sulphur lime stone filtration (PS)	Netherlands	65-70	> 90	Schippers et al., 1987
"Denitropur" process, four fixed-film upflow reactors (FS)	Monchenglabach Rasseln, Germany	75	95-100	CES, 1987
"Denipor" process, fixed-film, packed with floating polystyrene spheres (FS)	Langenfield & Monheim, Germany	65	95	CES, 1987
Down-flow filters (FS)	Neuss, Germany		92	CES, 1987
Fluidized bed experiment at two flows	Toulese, France	150	90	CES, 1987
Fixed-film granular bed, demonstration plant	Italy	40-50	80-90	Nurizzo et al., 1987
Upflow, fixed-film filters (PS)	Britain	NR	78-100	Richard and Faup, 1983
Fluidized-bed with varying methanol doses (PS)	Buklesham, Britain	62	100	Rittmann and Huck, 1989
Fluidized-bed (FS)	Stevenage, Britain	67	63	Hall et al., 1986

Process Description	Location	Influent Concentration NO_3/L	% Removal	Reference
Autotrophic denitrification (PS)	France	80	90-100	Blecon et al., 1983
Fixed-bed rotating biological disks (FS)	Laverence, California	60-80	91-93	Bouwer and Crow, 1988
Soil aquifer dune infiltration (FS)	Castricum, Netherlands	100	72	Piet and Zoeteman, 1985
Submerged up-flow filters (PS)	France	100-150	NR	Richard et al., 1980
Post-treatment using $FeCL_3$, following fixed film process to reduce effluent nitrite conc.	Germany	18 NO_2/l	100	Sontheimer et al., 1987
Packed-bed with polystyrene beads (PS)	France	55	95	Frank and Dott, 1985
Pseudomonas-packed columns (PS)	Europe	104	--	Nilsson and Ohlson, 1982
Substrate-cost evaluation in fluidized bed (PS)		--	--	Croll et al., 1985
Three stage process of oxygen removal, denitrification and reaeration (PS)	Germany	N.A.	--	Krutzenstein, 1982
Autotrophic, using hydrogen in fluidized bed reactor (PS)	Switzerland	N.A.	--	Kurt et al., 1987
Autotrophic, using columns packed with elemental sulphur and activated carbon (PS)	Germany	35	--	Overath et al., 1986
Autotrophic, using calcium alginate beads suspended in mixed reactor		120	--	Lewanaowski et al., 1987
Combined ion exchange biodenitrification (BS)	Netherlands	87-88	40-50	Van der Hock et al., 1988
Use of hydrogenotrophic denitrifiers in a polyurathane carrier reactor (PS)	Belgium	200-220	80	Dries et al., 1988
Bank filtration, soil aquifer (FS)	Rhine river, Germany	NR	75	Kubmaul, 1979

Note:
    PS = Pilot-scale study
    BS = Bench-scale study
    FS = Full-scale application
    NR = No report

5 *In Situ* Biodenitrification

In situ treatment is viable in situations where the contaminants (e.g., nitrates) are known and the extent of aquifer (or subsurface) contamination is well defined. With this restriction in mind, nitrate contamination in many areas throughout the United States is such that *in situ* biodenitrification might be very suited for nitrate removal. The primary reason for this suitability is the fact that there are many genera of denitrifying microorganisms of soil origin (Heinonen-Tanski and Airaksinen, 1988; Parrott, 1988; Soares et al., 1988); thus this process should not require the introduction of "nonnative" bacteria to a potential treatment site. In many situations, subsurface biodenitrification might be tremendously accelerated by simply providing the needed substrates (i.e., appropriate carbon sources) to carry out this process.

In situ denitrification is similar to conventional biodenitrification except that the process is carried out in the subsurface environment without having to pump the water out of the aquifer and without having to provide process equipment, tanks, and related physical plant appurtenances, instruments, and supplies. The apparent advantages to *in situ* denitrification, if the process can be made successful, include the following:

1. Considerable cost savings could be realized in plant capital investment for site development, pumping, piping, tankage, instrumentation, and other physical facilities.

2. Considerable savings in operating costs by reducing costs of energy used to operate treatment units and avoiding costs associated with pretreatment to avoid membrane fouling such as pre-chlorination and pH adjustment.

3. Significant savings in plant maintenance and replacement costs.

4. Significant cost savings in plant residuals disposal costs. Residuals include brines and regenerants from membrane systems and excess sludge from biodenitrification plants.

5. The need for highly trained operations and maintenance personnel is significantly reduced.

6. Subsurface biodenitrification appears to be more environmentally sound than other nitrogen removal systems since no waste products are generated and no ecological changes should occur if indigenous bacterial populations are utilized.

In situ biological denitrification is a fairly new field of research and relatively few successful studies have been reported (Janda et al., 1988; Mercado et al., 1988). However, experimental, pilot-plant and full-scale studies, conducted in Switzerland, Israel, Germany, England, France and the United States suggest that *in situ* treatment is a promising process. Several investigations have evaluated the introduction of various substrate and nutrients into aquifers in order to stimulate *in situ* denitrification. Substrates used include acetic acid, ethanol, and methanol as organic substrate as well as hydrogen and reduced sulphur as inorganic substrates. In many reported cases, phosphate was introduced as a microbial nutrient, while in autotrophic denitrification carbon dioxide or other inorganic carbon sources were used (Mercado et al. 1988; Janda et al., 1988).

Based on literature and conceptual evaluation, four alternative *in situ* denitrification methods are defined (Mercado, et. al., 1988):

Same well for recharge and pumping. In this alternative, a well is used for recharge of nitrate-polluted water with substrate. At a later stage, nitrate-free water is pumped out of the same well after time is allowed for sufficient biodenitrification to take place and the creation of a nitrate-free zone in the aquifer in the immediate area of the well. Generally, an additional well is needed to provide recharge water. Because this system is dependant on intermittent operation, several recharge and discharge wells are required to provide continuous operation.

Horizontal doublet system. This system utilizes two separate wells for injection and pumping at nominal distances ranging from 10 to 20 m. This system can supply nitrate-free water of constant quality on a continuous basis without the need for any surface treatment excluding disinfection. In this system, additional wells also are required to supply water for recirculation. As a minimum, a total three wells are needed. Experiments have shown that severe clogging due to gas and biomass can occur.

Vertical doublet system. This system consists of two wells of different depths, located in close vicinity to each other, or two wells constructed in one large diameter bore-hole. In this system, one well is used for injection of nitrate-contaminated water while the other is for the recovery of denitrified water. The basic difference between the horizontal scheme and the vertical scheme is that in the latter the biological reaction and the filtration zones are located vertically between the two wells. As in the previous case, an additional well also is needed to recirculate water through the system to accelerate the reaction rate. Also, as in the previous case, clogging problems can occur.

Daisy system. The daisy system consists of a production well surrounded by smaller diameter substrate injection wells. Organic substrate and nutrient are introduced into the aquifer through the injection wells by dilution using either treated water from the production well or from other nearby wells. The degree of nitrate reduction is dependant to great extent on geometrical factors such as the number and spacing of substrate injection wells, and the degree of lateral dispersion.

References

American Water Works Association Water (AWWA) Desalting and Reuse Committee (1989) Committee Report: Membrane Desalting Technologies. Journal AWWA 81:30

Andersen DR (1980) Nitrates in water supplies: Basis for concern and status of treatment processes. Department of Civil Engineering, University of Nebraska-Lincoln

Andersen DR (1981) Laboratory study of nitrate removal by ion-exchange, Department of Civil Engineering, University of Nebraska-Lincoln

Applegate LE (1984) Membrane separation processes. Chem Eng 91:64

Ashton M (1986) Reverse osmosis can be cost-effective. Public Works 117(8):55

Baier JH, Lykins Jr. BW, Frank CA, Kramer SJ (1987) Using reverse osmosis to remove agricultural chemicals from ground water. Journal AWWA 79(8):55

Blecon G et al. (1983) Procede de denitrification biologique autotrophe par thiobacillus denitrifications of sulphur maerl. Revue Francaises des Sciences de l'Eau 2(3):267

Bouwer EJ, Crow PB (1988) Biological processes in drinking water treatment. Journal AWWA 80(9):84

Buros OK (1989) Desalting practices in the United States. Journal AWWA 81:38

Clifford D (1987) Nitrate removal from drinking water in Glendale, Arizona. U.S. EPA Report EPA/600/S2-86/107, March

Consultants in Environmental Sciences (1987) Effects of nitrate removal on water quality in distribution. Final Report. Contract PECD 7/7/218. Department of the Environment, UK

Croll BT et al. (1985) Biological fluidized bed denitrification for potable water. In: Tebbut THY (ed) Advances in Water Eng. Proceedings of the International Sym. Elsevier Appl. Sci., London, p 180

Dahab MF (1987) Treatment alternatives for nitrate contaminated groundwater supplies. Journal of Environmental Systems 17:65-75

Dahab MF, Lee YW (1988) Nitrate removal from water supplies using biological denitrification. Journal Water Pollution Control Federation (WPCF) 60(9):1670

Dahab MF (1985) Potential for nitrate removal using biological denitrification. Proc. ASCE National Conference on Environmental Engineering, Boston, MA, July

Dore M, Simon PH, Deguin A, Victor J (1986) Removal of nitrate in drinking water by ion exchange--Impact on the chemical quality of treated water. Water Research 20(2):221

Dries D, Liessens J, Verstrate W, de Vos P, de Ley J (1988) Nitrate removal from drinking water by means of hydrogenotrophic denitrifiers in a polyurethane carrier reactor. Water Supply 6:181-192

Dykes GM, Conlon WJ (1989) Use of membrane technology in Florida. Journal AWWA 81:43

Eisenberg TN, Middlebrooks EJ (1984) A survey of problems with reverse osmosis water treatment. Journal AWWA 76(8):44

Frank C, Dott W (1985) Nitrate removal from drinking water by biological denitrification. Vom Wasser 65:287

Gayle BP, Boardman GD, Sherrard JH, Benoit RE (1989) Biological denitrification of water. Journal of Environmental Engineering 115: 5

Guantlet RB (1981) Removal of ammonia and nitrate in the treatment of potable water. Water Research Centre, Medmenham Laboratory, Publishers Chechester

Guter GA (1990) Personal communication, July

Guter GA, Kartinen EO (1989) Alternatives for reducing nitrate in municipal water supplies. Presented at the Symposium on Nitrates in Drinking Water, American Chemical Society, Dallas, TX, April

Guter GA, Lauch RP (1988) Nitrate treatment in McFarland: Final study results and costs. Presented at the 1988 Annual Conference of the AWWA, Orlando, FL, June

Guter GA (1987) Nitrate removal from contaminated water supplies: Volume I. Design and initial performance of a nitrate removal plant, U.S. EPA Report EPA/600/S2-86/115, April

Guter GA (1982) Removal of nitrate from contaminated water supplies for public use: Final report, EPA-600/2-82-042. August

Hagen K, Holl W, Kretzchmar W (1986) The CARIX process for removing nitrate, sulphate and hardness from water. Aqua 5:275

Hall T, Walker RA, Zabel TF (1986) Nitrate removal from drinking water - Preliminary guide to process selection and design. WRC Processes, International Rept. 422-S., Stevenge, England

Heinonen-Tanski H, Airaksinen AK (1988) The bacterial denitrification in ground water. Water Science and Technology 20(3):225

Janda V, Rudovsky J. Wanner and K. Marha (1988) In-situ denitrification of drinking water. Water Science and Technology 20(3):215

Kreevoy MM, Nitsche CI (1982) Progress toward a solution to the nitrate problem. Env Sci and Tech 16:9

Kruithof JC, van Bennekom CA, Dierx HAL, Hijnen WAM, van Paassen JAM, Schippers JC (1988) Nitrate removal from ground water by sulphur/limestone filtration. Water Supply 6:207

Krutzenstein K (1982) Apparatus and methods for denitrifying water. Patent Application 3121395, West Germany

Kubmaul H (1979) Purifying action of the ground in the treatment of drinking water: Oxidation techniques in drinking water treatment. W. Kuhn and H. Sontheimer (eds) EPA 570/9-79-020

Kurt M, Dunn IJ, Bourne JR (1987) Biological denitrification of drinking water using autotrophic organisms with hydrogen in a fluidized bed biofilm reactor. Biotechnology and Bioengineering 29:493

Lauch RP, Guter GA (1986) Ion exchange for the removal of nitrate from well water. Journal of the American Water Works Association 78:83-88

Lee YW (1987) Evaluation of packed-bed bio-denitrification for nitrate removal from water supply. Master of Science thesis, University of Nebraska Libraries, Lincoln, NE.

Lewanaowski Z, Bakke R, Characklis WG (1987) Nitrification and autotrophic denitrification in calcium alginate beads. Water Science and Technology 19:175

Marquardt OK (1987) Reverse osmosis process for removing nitrate from water. Aqua 1:39-44

Mercado A, Libhaber M, Soares MIM (1988) In-situ biological groundwater denitrification, concepts and preliminary field tests. Water Science and Technology 20(3):197

Metcalf and Eddy, Inc. (1979) Wastewater Engineering: Collection, Treatment, Reuse. 2nd edition, McGraw-Hill

Montgomery JM Consulting Engineers, Inc. (1985) Water Treatment, Principles & Design. John Wiley and Sons

Nilsson I, Ohlson S (1982) Columnar denitrification of water by immobilized *pseudomonas denitrificans* cells. European Journal of Applied Microbiology and Biotechnology 14:86

Nurizzo C, Vismara R, Allione MR (1987) Potable waters biodenitrification fixed film demonstration plant using sugar as organic carbon source. R & D Dept. Castagnetti SPA

Overath H, Hussaman A, Haberer K (1986) Biological nitrate removal by *thiobacillus denitrificans* using elemental sulphur fixed on activated carbon electron donor. Vom Wasser 65:59

Parrott JD (1988) Shallow groundwater denitrification associated with an oxidation-reduction zone in Hall County, Nebraska. Master of Science Thesis, University of Nebraska Libraries, Lincoln, NE

Philipot JM, De Larminat G (1988) Nitrate removal by ion-exchange: the Ecodenit process at industrial scale in Binic (France). Water Supply 6(1):45-50

Philipot JM (1985) Biological techniques used in the preparation of drinking water: Nitrate, Iron, and Manganese removal. EPA Research Symposium, Cincinnati, August

Philipot JM (1982) Denitrification des eaux de consummation, Une realisation en course en region paisienne: Eragny. L'Eau, L'Industrie, Les Nuisance 69:27

Piet GJ, Zoeteman BCJ (1985) Bank and dune filtration of surface water in the Netherlands. In Asano T (ed) Artificial Recharge of Ground Water. Butterworth Publication, Boston, MA

Rautenbach R, Kopp W, Hellekes R, Peters R, Van Opbergen G (1986) Separation of Nitrate from well water by membrane processes (reverse osmosis/electrodialysis reversal). Aqua 5:279-282

Ravarini P, De Larminat G, Couttelle J, Bourbigot MN, Rogalla R (1989) Large scale biological nitrate and ammonia removal. The Institute of Water and Environmental Management (IWEM) Conference, September

Richard YR (1989) Operating experience of full scale biological and Ion-exchange denitrification plante in France, Institute of water and environment management (IWEM) 88 Conference paper

Richard YR, Burriat J (1988) Azurion elimination des nitrates par resines echangeuses d'ions. TSM L'Eau, Avril, 227-230

Richard Y, Faup GM (1983) The removal of nitrogen by fixed cultures in upflow beds. In: Urbistondo R, Trueb E (eds) World Water Supply 1:2/3: SS-1-SS7-4

Richard Y, Leprince A, Martin G, Leblanc C (1980) Denitrification of water for human consumption. Water Science and Technology 12:173

Rittmann BE, Huck PM (1989) Biological treatment of public water supplies. Critical Reviews in Environmental Control 19:2

Schippers JC, Kruithof JC, Mulder FG, van Lieshout JW (1987) Removal of nitrates by slow sulphur limestone filtration. Aqua 5:274

Sharefkin J, Shechter M, Kneese A (1985) Impacts, costs and techniques for mitigation of contaminated groundwater. Water Resources Research 20(12):1771-1783

Soares MIM, Belkin S, Abeliovich A (1988) Biological groundwater denitrification: laboratory studies. Water Science and Technology 20(3):189

Sontheimer H et al. (1987) Biological denitrification of water with minimum post-treatment. Patent Application 3603123 West Germany

nitrogen removal methods. Different techniques based on physico-chemical or biological principles were investigated and compared for technical and economical feasibility (Germonpre, 1988). Membrane separation is possible (Rautenbach et al., 1986; Hellekes et al., 1989), and has been demonstrated (Kneifel et al., 1988), but costs have been prohibitive for full-scale applications for nitrate removal alone (Gros et al., 1982).

This paper summarizes the results obtained in full-scale plants removing nitrate from drinking water. Operational results are available from biological and ion-exchange plants.

2 Ion Exchange

Ion exchange is a widely applied process for industrial-water treatment, but health considerations regarding the release of undesirable substances by the synthetic resins had initially slowed the application of this technology (Philipot et al., 1982). After intensive pilot investigations, improvements in resin quality and specificity led to full-scale applications starting in the United States in 1983 (Lauch et al., 1986) and leading to the largest plant for nitrate removal in France (Richard, 1989). The nitrate eliminated from the water is concentrated into a brine effluent containing regeneration salt. Biological treatment has been proposed for its elimination (Van de Hoek et al., 1988), and the discharge to sewers is widely applied (Dumbleton, 1990; Philipot, 1988; De Larminat et al., 1990).

The treatment plant at Binic in the western region of France (160 m^3/h) has extensive operating experience (Godart et al., 1991). This plant treats surface water with a variable nitrate concentration between 45 and 100 mg/L, with peak values up to 165 mg NO_3^-/L. The resin beds are located behind a conventional coagulation/settling/filtration unit and before final ozonation. The nitrate is continuously analyzed by UV-spectroscopy. Fig. 1 shows the evolution of incoming and outflowing nitrate, with an average removal rate of 75%. A typical analysis of inlet and outlet of the plant is given in Table 1. Since the resins also remove other ions, about 90% of sulfate ions and 30% of carbonates are eliminated, whereas the water is enriched in chlorides.

Three ion exchangers of 2700 L of resins are alternated with operation and regeneration cycles, one cycle being at least 5.5 hours at peak flows. The production cycle is finished

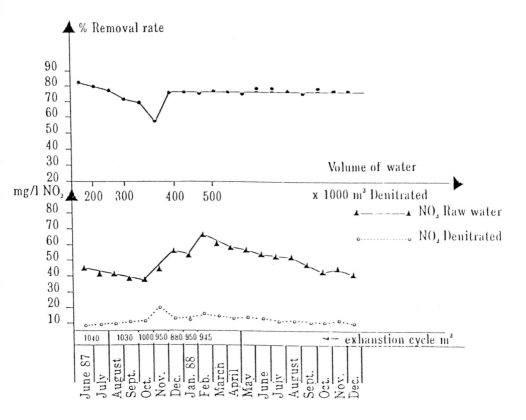

Fig. 1. Binic inflow and outflow nitrate concentration

Table 1. Ion exchange plant at Binic: Water quality

Compound (mg/L)	Raw Water	Treated Water
NO_3^-	70	25
SO_4	50	10
HCO_3	65	55
Cl	50	110

when the nitrate leakage is around 10 mg NO_3/L. About 0.3 kg of NaCl/m³ of denitrified water is used for regeneration. A typical cycle is shown in Fig. 2; resin exhaustion is reached after a volume-to-volume ratio of 400 m³ of water per m³ of resin bed is reached. In Binic, the wastes are sent to a sewage works in the vicinity of the denitrification plant which accepts 55 m³/d of resin regeneration liquor, representing 450 kg of NO_3^- and 516 kg of Cl⁻, or about 20% of the nitrogen loading of the wastewater treatment plant (Philipot et al., 1988). The latter achieves complete removal of nitrogen with residual below 5 mg total N/L.

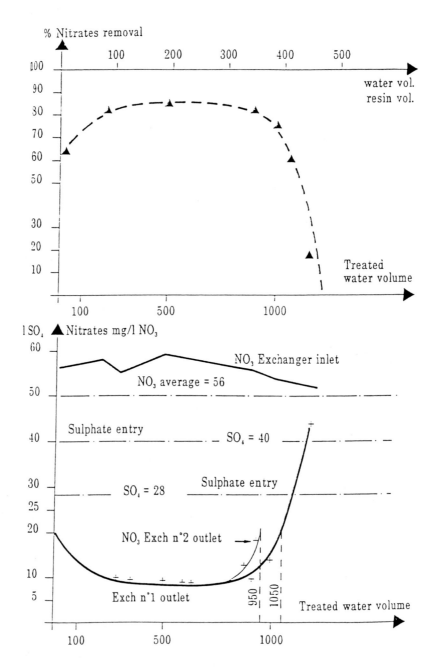

Fig. 2. Binic denitration: Observation of exhaustion cycle

After thirty months of operation of the Dowex SBRP resins, the cycle times had to be shortened by 10% because of fouling by organic matter. These compounds from the river water, which was highly exposed to pollution, were unsatisfactorily removed by the upstream pretreatment facility. Chemical washing did not restore the nitrate-removal efficiency, and the resins were changed after thirty-six months. Several measures were implemented to increase the bed life of the ion exchanger: (1) thicker sand beds on upstream filters to further reduce organics, (2) more intense backwashing, and (3) double regeneration every four cycles.

3 Biological Nitrate Removal

Biological treatment of drinking water to remove a variety of compounds (Rapinat, 1982) has recently increased in popularity because of its low cost and reduced health and taste impacts (Bouwer et al., 1988). Biological methods may be easily adapted to conventional drinking-water technologies by the use of fixed bacteria on filter grains (Sibony, 1982). In addition, especially for surface water with high organic and ammonia concentrations, biological sand or activated carbon filters reduce odorous substances (Bourbigot et al., 1986).

3.1 Autotrophic Processes

Autotrophic bacteria do not require an outside carbon source to degrade nitrate (Kurt et al., 1987), but their reaction rate is low because of the slow growth rate of autotrophic organisms. On the other hand, sludge production is much lower. Their attachment on surfaces led to the full-scale application of this process (Gros et al., 1986). Studies using fluidized carriers for intensified reactions are under way (Dries et al., 1988; Tuisel et al., 1989). Table 2 lists a few experiments on autotrophic denitrification through the oxidation of sulfur or hydrogen.

To avoid handling the hydrogen gas, a sulfur limestone process was tested (Le Cloirec et al., 1988) and proved to be a simple process to operate, although clogging of gas bubbles required vacuum deaeration and upflow filtration. The slow removal rates led to the implementation of a large subsoil reactor (Kruithof et al., 1987), followed by cascade aeration and infiltration pond for polishing. This process has been in operation for three years, but the reactor bed had to be deepened after one year to avoid nitrate breakthrough.

from a highly cultivated plain and partly by riverbank infiltration of the Seine. The concentrations of nitrate and ammonia were close to the EC Guidelines and were increasing. The absence of an alternative source made it necessary to provide treatment in anticipation of further ground-water nitrogen enrichment.

A flow scheme based on earlier studies (Rogalla et al., 1990) is shown in Fig. 3. The first reactor is packed with a mineral medium, heat-expanded shale. The principle of the BIODENIT process is based on conventional sand filters: the water flows downward under slight pressure on a mineral medium. The heterotrophic bacteria attach to the granular material because of its high immobilizing characteristics: large specific surface and high macroporosity. This material has a low density and a good resistance to abrasion. Grain sizes between 2 and 5 mm were selected to favor bacterial adherence and limit head loss.

To obtain balanced biological growth, phosphorus must be added in addition to the carbon source for the denitrifiers. The carbon substrate is ethanol, a product that bacteria can metabolize but which is not toxic for human consumption. A polishing treatment is required downstream from the denitrifying filter since the BIODENIT effluent contains no dissolved oxygen and the bacterial metabolisms generate easily biodegradable organic carbon. The water is polished on an aerated two-layer sand and activated carbon filter before ozonation. Table 3 shows average inlet and outflow concentrations.

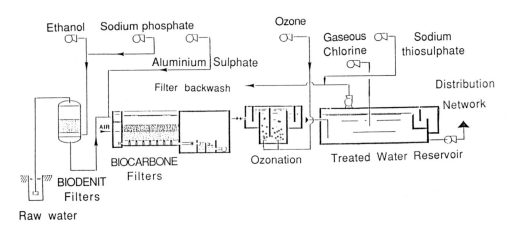

Fig. 3. Flow diagram of Dennemont plant

Table 3. Water quality at the Guernes/Dennemont plant

Quality	Raw Water	Treated Water
Temperature (°C)	12/13	-
pH	7.2/7.4	7.5/7.6
Nitrates (mg/L)	40/65	15/17
Nitrites (mg/L)	0.1	0
Ammonium (mg/L)	2/3.5	0.01/0.02
Turbidity (NTU)	0.3	0.2
Phosphates (mg/L)		0.1
Orthophosphates (mg/L)	0.1	-
TOC (mg/L)	1.3	1
BDOC (mg/L)	< 0.1	< 0.5
Total aluminium (μg/L)	-	< 20
Total alkalinity (mg Ca/L)	300	295

Biological two-layer filtration for micropollutant and ammonia removal is more and more common on a large-scale basis without aeration (Bablon et al., 1988), but oxygen becomes a limiting factor at higher substrate concentrations. Air is injected into the middle of the filter at the bottom of the carbon layer, and coagulant can be dosed into the filter feed to increase solids retention. Treated water is stored and used for backwashing. Both filters use the same backwashing equipment for air expansion and water rinsing. The backwashing can be fully automated with pneumatic valves.

The unit was commissioned at the end of 1987. A daily sludge production of 32 kg/d is discharged into the sewer. The raw and treated water characteristics are shown in Table 3. The calco-carbonic equilibrium is not influenced markedly, since bicarbonate is produced through denitrification and consumed by nitrifiers. Phosphorus and aluminum levels are not significantly altered. Ethanol has never been detected at the outflow of the treatment plant, since a considerable safety margin for removal of excess carbon is provided by the aerated filters.

Micropollutants are reduced by the biological system as shown in Table 4. Several mechanisms explain the removal of organic pollutants: anoxic biological uptake (Bouwer et al., 1988), aerobic degradation of toxic compounds (Kantardjieff et al., 1990), stripping in the aerated filters (Philipot et al., 1985), and activated carbon adsorption (Graese et al., 1987). The latter may be enhanced by bioregeneration of the aerated filters.

5 Economic Aspects

Investment in denitrification can create problems for utilities which have not prepared for the inevitable added costs of ground-water treatment. Table 5 shows the operating costs expressed per cubic meter of ground water treated for both ion exchange and heterotrophic denitrification. This cost does not include waste treatment, which is normally provided through sewer discharge. The operating cost of a biological denitrification unit is comparable to an ion-exchange facility or to a conventional surface water clarification facility. Investment costs are slightly higher for biological reactors, making them a more viable option for larger plants.

Table 5. Operating cost of full-scale facilities in France in FF/m^3 (1989)

Factor	Biological	Ion Exchange
Plant flow m^3/h	400	160
Reagents consumption	0.5	0.6
Electricity	0.1	0.1
Personnel (2 h of supervision/d)	0.2	0.1
Maintenance and renewal	0.3	0.2
Payback	0.5	0.3
Total	1.6	1.3
In US $/1000 Gal.		
Operation and Maintenance	0.84	0.76
Payback	0.38	0.23
Total	1.2	1.0

6 Conclusions

After testing a large number of alternatives, the technical and economical feasibility of both the heterotrophic biological process and the ion-exchange nitrate-removal process were confirmed. More than ten plants are operating in France. Presently, these two methods are by far the most economical with regard to investment costs. Ion exchange resins were less durable than initially expected on surface water with a high carbon concentration. Special methods have been applied to control the biological development and efficiency of the fixed bacteria. Total treatment cost is between 1 and 1.5 FF/m^3 (0.75 to 1.2 US $/1000 Gal).

Acknowledgements

Numerous advisers and technical helpers were involved in the development and implementation of the biological denitrification plants, whose contributions are appreciated. Mr. Sibony and Philipot led the early development of the processes. On-site data were collected by Mr. Sanson, Mr. Marteil and Mrs. Ducray. The authors are grateful for all these efforts.

References

Bablon G, Ventresque C, Ben Aim R (1988) Developing a sand-GAC filter to achieve high-rate biological filtration. AWWA 80(12):47-53

Bourbigot MM, Sibony J (1986) Providing water for the Ile de France. JAWWA 78(7):62-71

Bouwer EJ, Crowe PB (1988) Biological processes in drinking water treatment. JAWWA 80(9):82-93

Bouwer EJ, Wright JP (1988) Transformation of trace halogenated aliphatics in anoxic biofilm columns. Journal of Contaminant Hydrology 2:155-169

Braester C, Martinell R (1990) Theoretical and practical aspects of vyredox and nitredox plants. In: Modern methods of water treatment. Int Symposium April 22-24, 1990. Czechoslovakia, Pribram

Bremner J, Shaw K. (1958) Denitrification in soils. J Agr Sc 51:39-52

Christensen, Henze M, Harremoes P (1977) Biological denitrification of sewage - a literature review. Prog Wat Tech 8(4/5):509-557

Cooper PF, Wheeldon DHV (1980) Fluidized and expanded bed reactors for wastewater treatment. Wat Pol Control 79:286-306

Dahab MF, Lee YW (1988) Nitrate removal from water supplies using biological denitrification. JWPCF 60 9:1670-1674

De Larminat G, Deboves JJ, Cleret D (1990) Une unité de dénitratation pour la ville de Sainte Adresse. L'eau, l'Industrie, les Nuisances 135:53-56

De Constantin S, Pascal O, Block JC, Dollard MA (1986) Bacterial population in distribution networks: comparison of two full scale studies. Wat Supply 4:199-215

Dillon G, Thomas V (1989) A pilot evaluation of the Biocarbone Process for settled sewage treatment and tertiary nitrification of secondary effluent. In: Wat Sci Tech 22:1/2

Dries D, Liessens J, Verstraete W, Stevens P, De Vos P, de Ley J (1988) Nitrate removal from drinking water by means of hydrogenotrophic denitrifiers with polyurethane carrier. Wat Supply 6:181-192

Dumbleton, B (1990) Ioning out high nitrate levels. In: World Wat and Environ Eng June 1990 45-46

Eppler D, Eppler A (1986) Neues Verfahren zur biologischen Denitrifikation von Grundwasser. Wasserwirtschaft 76 11:492-494

European Community Council directive of July 1980 relating to the quality of water intended for human consumption. 80/778/EEC Off J Eur 88 Commun.23 L 229 11-29 (1980)

Gauntlett RB, Craft DG (1979) Biological removal of nitrate from river water. In: Technical Report TR 98, WRC, Medmenham, England

Germonpre R (1988) Different possibilities for the elimination of nitrates in surface waters. Wat Supply 6:63-70

382

Gilles P, Bourdon Y (1985) Nitrification and denitrification with fixed bacteria.(in French). L'Eau, l'Industrie, les Nuisances 93:53-57

Godart H, Gonnard P (1991) Co-current denitration on resin :three years practical experience. In: IWSA Copenhagen Congress 1991 Special subject n° 1

Graese SL, Snoeyink VL, Lee RG (1987) Granular activated carbon filter adsorber systems. JAWWA 79(12):64-74

Gros H, Ginocchio JC (1982) Drinking water denitrification :pilot studies of three processes. In: IWSA Conference, Zürich, SS 7, 24-26

Gros H, Schnoor G, Rutten G (1986) Nitrate removal from groundwater by autotrophic microorganisms. Wat Supply 4:11-21

Hambsch, B, Werner, P (1989) Optimization and control of biological denitrification processes by measuring the growth rate of bacteria (in German). Vom Wasser 72:235-247

Harremoes P, Jansen JLC, Kristensen GH (1980) Practical problems related to nitrogen bubble formation in fixed film reactors. Prog Wat Tech 12:253-269

Hellekes R, Van Opbergen G, Kopp W, Rautenbach R (1989) Nitratelimination durch Umkehrosmose und/oder Elektrodialyse Verfahrensentwicklung und Verfahrensoptimierung (in German). Wasser-Abwasser 12:638-844

Janda V, Rudovsky J, Wanner J, Marha K (1988) In situ denitrification of drinking water. Wat Sci Tech 20:215-219.

Jestin JM, Philipot JM, Berdou C, Moulinot JP (1986) Control of a biological process: denitrification at Eragny (in French). TSM - L'eau 81 7-8:359-362.

Joret JC, Levi Y (1986) Rapidly evaluating assimilable carbon. Trib Cebedeau, 510 39:3-9

Joret JC, Levi Y, Paillard H, Ravarini P (1989) Comparison of different full-scale drinking water treatment plants for removal of biodegradable dissolved organic carbon. AWWA Annual Conference, 1989, Los Angeles

Kantardjieff A, Jones J P, Zaloum R, Araneda A (1990) Degradation of toxic compounds in TMP mill effluents by biological aerated filter, TAPPI Environmental Conference Seattle April, 251-259

Kneifel K, Lührs G, Wagner H(1988) Nitrate removal by electrodialysis for brewing water. Desalination 68:203-209.

Koopman B, Stevens CM, Wonderlick CA (1990) Denitrification in a moving bed upflow sand filter. Research JWPCF 62:3:239-245

Kruithof JC, Schippers JC, Mulder FG, Van Lieshout JW (1987) Nitrate removal by Slow sulphur/limestone Filtration. Aqua 5:274-280

Kurt M, Dunn J, Bourne JR (1987) Biological denitrification of drinking water using autotrophic organisms with H2 in a fluidized bed biofilm reactor. Biotechnology and Bioengineering 29:493-501

Lauch RP, Guter GA (1986) Ion exchange for the removal of nitrate from well water. AWWA 78(5):83-86

Le Cloirec P, Martin G (1988) Biological autotrophic process for removing total inorganic nitrogen. Wat Supply 6:151-156

MacDonald DV (1989) Denitrification by fluidized biofilm reactor. Research Journal WPCF 62:796-802

Makita N, Fuchu Y, Kimura H (1988) Biotreatment of lake water by biological aerated filter. 1st Particulate Technology Conference, Sept 1988, Malaysia

Mercado A, Libhaher M, Soares MIM (1988) In situ biological groundwater denitrification: concepts and preliminary field tests. Wat Sci Tech 20:197-209

Nilsson I, Ohlson S (1982) Columnar denitrification of water by immobilized Pseudomonas denitrificans cells. Europ J Appl Microb Biotechnol 14:86-90.

Nurizzo C, Vismara R (1988) Biological Denitrification: Pilot plant results using sugar as organic carbon source (in Italian). Ing Ambien 17:2:88-95

Philipot JM, Patte A (1982) Water denitrification by a biological process (in French). Techniques et Sciences Municipales - L'eau, 77 4:165-172

Philipot JM, Chaffange F, Pascal O (1985) Biological denitrification: A year's summary at the Eragny plant (in French). Wat Supply 3(1):93-98.

Philipot JM, Sibony J (1985) Elimination of chlorinated solvents, Wat Supply 3:203-210

Philipot JM, De Larminat G (1988) Nitrate removal by ion exchange: the Ecodenit process at industrial scale in Binic (France). Wat Supply 6(1):45-50

Polprasert C, Park HS (1986) Effluent denitrification with anaerobic filters. Wat Res 20(8):1015-1021

Rapinat M (1982) Recent developments in water treatment in France. AWWA 74(12):610-617

Rautenbach R, Kopp W, Hellekes R, Peters T, Van Opbergen G (1986) Separation of nitrate from well water by membrane processes (RO/EDR). Aqua 5:279-282

Richard YR (1989) Operating experiences of full scale biological and ion exchange denitrification plants in France. J IWEM 3(2):154-167

Roennefahrt KW (1986) Nitrate elimination with heterotrophic aquatic microorganisms in fixed bed systems with buoyant carriers. Aqua 5:283-285

Rogalla F, Bacquet G, Payraudeau M, Bourbigot MM, Sibony J, Gilles P (1989) Nitrification and P-precipitation with biological aerated filters. Res. JWPCF 62:169-176

Rogalla F, Ravarini P, Coutelle J, Damez F (1990) Large scale biological nitrate and ammonia removal, JIWEM 4:319-329

Rudd Th (1987) Effects of nitrate removal on water quality in distribution. Final Report of contract PECD 7/7/218 to UK Dept of Environment, London

Seropian JC, Vergne C, Moro A, Capdeville B (1990) Water denitrification: high efficiency biological fluidized bed. Wat Sci Tech 22 1/2 (Conference on Biofilm Reactors, Nice)

Servais P, Billen G, Hascoet MC (1987) Determination of the biodegradable fraction of dissolved organic matter in water. Wat Res 21(4):445-450

Sibony J (1982) Development of aerated biological filters for the treatment of waste and potable water. In: IWSA Conference, Zurich, SS 9, 25- 31

Soares MIM, Belkin S, Abeliovich A (1989) Clogging of microbial denitrification sand columns: gas bubbles or biomass accumulation. Wasser-Abwasser 22:20-24.

Tuisel H, Heinzle E, Luttenberger H (1989) Biological Denitrification of Drinking Water with hydrogen in a fluidised bed reactor (in German) GWF Wasser Abwasser 130:10-13

Van der Hoek JP, Van der Ven M, Klapwijk A (1988) Combined ion exchange/biological denitrification for nitrate removal from groundwater. Wat Res 22:679-684

Van der Hoek JP, Zwanikken, B, Griffioen AB, Klapwijk A (1988) Modelling and Optimisation of Ion Exchange and Biological Denitrification. Z Wa Abw Fors 3:85-91

A NEWLY DEVELOPED PROCESS FOR NITRATE REMOVAL FROM DRINKING WATER

A.F. Miquel and M. Oldani[1]
Otto Oeko-Tech GmbH & Co KG
Gustav Heinemann Ufer 54
D-5000 Köln, GERMANY

Abstract

A newly developed process to remove nitrate from drinking water is discussed. It is concluded that selective electrodialysis is a technically and economically attractive alternative for denitrification of drinking water. A key factor in this respect is the quality and composition of the treated water. The fact that the separation can be done without using chemicals makes this process especially attractive. Because of the composition of the wastewater, it may be treated biologically to eliminate the nitrates, thus making the process comparable to direct biological treatment of the drinking water. Operation and maintenance of an electrodialysis plant is very simple and fully automated, making the process applicable to pumping stations of any size.

1 Introduction

Nitrogen is an important building block in nature's chemistry. The atmosphere we breath contains 78 vol% of molecular nitrogen, and nitrogen is also a key component in the structure of proteins and amino acids without which life would not exist.

The natural nitrogen contents of the soil are of organic origin, and have entered the soil through natural cycles and processes. Organic and inorganic nitrogen concentrations in soil, ground water, and estuaries have increased significantly as a result of human intervention in this natural nitrogen cycle. One of the major sources of nitrogen pollution in our water and soil environment is agriculture.

Because nitrate has been linked to a number of health hazards, as described by other contributors to this volume, limits have been established for nitrate concentrations in drinking

---

[1]Asea Brown Boveri Ltd., Corporate Research, CH-5405 Baden, SWITZERLAND

NATO ASI Series, Vol. G 30
Nitrate Contamination
Edited by I. Bogárdi and R. D. Kuzelka
© Springer-Verlag Berlin Heidelberg 1991

water. One method of meeting these limits is to treat the water to reduce nitrate concentrations. This paper discusses a novel treatment method and compares it to other common methods.

2 Primary and Secondary Solutions

A long-range program to reduce nitrate levels in ground and/or surface water would require extensive changes in industrial, agricultural and household behavior. Such a long-range program would begin with the reduction of sources such as acid rain, go on to encompass a full extension of wastewater treatment to include denitrification, and end with a redirection of agricultural policies away from the present effort to maximize yields. These "primary" measures should certainly be encouraged and implemented as quickly as possible. However, it is uncertain that the necessary policy measures could be implemented rapidly because of opposition from different interest groups. Second, even if the necessary measures could be implemented, it is questionable whether they would lead to an immediate improvement in the situation because of the large storage capacity for nitrates in the soil and the long residence times of water in the ground. In the meantime, technical solutions are required in order to meet drinking-water regulations in the affected areas.

2.1 Water Blending

The blending of nitrate-contaminated water sources with "clean" water is usually one of the first measures taken to reduce nitrate concentrations in delivered drinking water. Typically water from outside the water supply system has to be purchased. This extension or interlinking of water supplies thus has the welcome side effect of providing a higher supply security for the system. However, the approach usually requires large capital expenditures to build new piping systems, pumping stations and control systems. Furthermore, mixing of water with possibly very different compositions may require taking additional treatment measures in order to avoid corrosion in the mains.

2.2 Treatment Processes

Water-treatment technologies aimed at removing nitrate from drinking water also have been developed. Processes that are able to remove nitrate from a water stream fall within the following classes:

1. reverse osmosis

2. ion-exchange processes

3. anaerobic biological digesters

4. electrodialysis

The suitability of these processes for nitrate removal is analyzed below.

2.2.1 Reverse Osmosis

Reverse osmosis, like most membrane separation processes, relies on pressure to drive water through a membrane. In this case the membrane pores must be small enough to reject the solvatized salt ions, thus achieving a salt-removal effect. The applied pressure is at least as large as the osmotic pressure difference across the membrane, plus the pressure required to actually force a significant amount of water through the membrane.

Because the filtrate or permeate of this process contains too few ions, (i.e., it approaches the quality of distilled water), it has to be mixed with untreated water in order to achieve a drinkable composition. This fact limits the applications to nitrate reductions of approximately 50%; the treated water will then have roughly 50% of the original concentration of ions.

Acid has to be added to the feed stream in order to avoid scaling from hard water (the acid is then discharged with the brine stream). The maximum achievable water recoveries are in the region of 70 to 80%; that is, 20 to 30% of the feed water is discharged as a brine. Disposal of the wastewater poses a number of problems since the brine stream contains a high concentration of salts from the water itself as well as from the added acid. Further pretreatment of the feed stream is usually necessary, comprising such steps as a fine filtration and removal of iron and manganese. The drinking water will also require treatment. Since the total salt content is reduced by roughly the same portion as the nitrate, either the addition of calcium hydrate or passing the water through a dolomite filter will be required. A pH adjustment, meaning a removal of excess carbon dioxide or an addition of acid to the treated water, will be necessary in most cases as well. Finally, since membranes are exposed to biological fouling, especially on the filtrate side after stand-still periods, a disinfection stage is also necessary.

over 30 years, generally for the production of drinking water from brackish and sea water (McRae, 1988). In Japan it is also used to concentrate seawater for table-salt production (McRae, 1988). The driving force for the process is an electrical DC field which forces the ions to migrate. This migration is limited by the introduction of charge-selective membranes: anions can only pass the anion-exchange membrane, cations the cation-exchange membrane (Fig. 1). An alternating arrangement of the two different membranes splits the feed-water stream in two, with one stream containing demineralized water and another containing the ions removed from the diluate (Fig. 2). The brine stream is generally recycled back to the stack in order to achieve higher recovery rates (Fig. 2).

Electrodialysis can be compared to reverse osmosis as a denitrification technique (Rohmann et al., 1985). The process primarily desalinates the water stream; nitrate removal is a side effect. Conventional electrodialysis can, however, achieve the same nitrate-removal rate as reverse osmosis while requiring less acid dosing and reaching higher water-recovery rates (Rautenbach, et al., 1987). Thus, smaller quantities of brine carrying smaller salt loads are produced by electrodialysis than by a comparable reverse osmosis. Conventional electro-dialysis and reverse osmosis have similar specific production costs (Rautenbach et al., 1987).

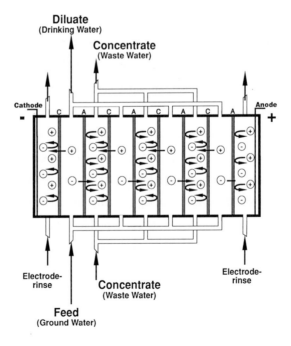

Fig. 1. The principle of electrodialysis

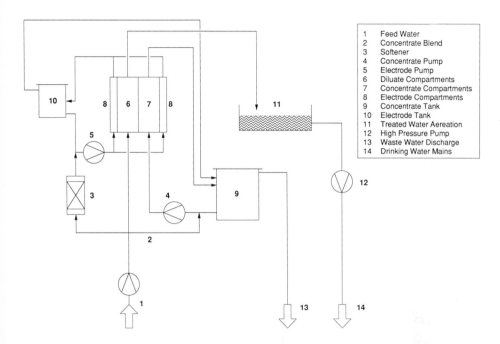

1	Feed Water
2	Concentrate Blend
3	Softener
4	Concentrate Pump
5	Electrode Pump
6	Diluate Compartments
7	Concentrate Compartments
8	Electrode Compartments
9	Concentrate Tank
10	Electrode Tank
11	Treated Water Aereation
12	High Pressure Pump
13	Waste Water Discharge
14	Drinking Water Mains

Fig. 2. Typical flow sheet of an electrodialysis plant

3 Selective Nitrate Removal by Electrodialysis: NITREM®

Based on the analysis presented above, electrodialysis was identified as the technology with the highest development potential for this application. The aim was to turn electrodialysis into a selective nitrate-removal process, thereby reducing overall water desalination, chemical dosing and the amount and salt load of the wastewater.

The key aspect in reaching selectivity for a specific towards any kind of ion is membrane selection. A number of factors influence membrane selectivity, foremost of which is membrane/ion affinity. However, any type of potential membrane selectivity towards a single ion or a class of ions can be nullified if the right operating conditions are not present. For this reason, a selective membrane should be operated as near as possible to the thermo-dynamic equilibrium state, and any type of transport phenomena influences should be avoided. Any potentially selective membrane has thus to be operated at low current densities and at high flow velocities. (There are of course limits to the range in one can vary these parameters; such limits are usually set by cost concerns; i.e., cost concerns dictate that membrane surface area and pressure drop through the stack should be kept to a minimum.)

The research carried out at the Asea Brown Boveri Ltd. (ABB) Research Center yielded a selective nitrate-removal process based on the electrodialysis process that is able to remove a maximum amount of nitrate and as little as possible of other ions from the drinking-water stream. Furthermore, a high water-recovery ratio (purified drinking-water quantity in relation to the wastewater) was achieved. Thus, water is produced containing all the original components desired and required by the human organism and fulfilling the limits set for nitrate concentrations by the authorities. Pilot-scale experiments in the ABB Research Center and at two Swiss communities have proven the effectiveness of this nitrate-selective removal process (NitRem®) under actual operating conditions (Fig. 3.). The NitRem® process can produce drinking water with less than 25 mg/L nitrate even when nitrate concentrations in the drinking water increase, as is expected for the near future. In addition, a certain water-softening effect is inherent to the process (Table 1), a welcome side-effect.

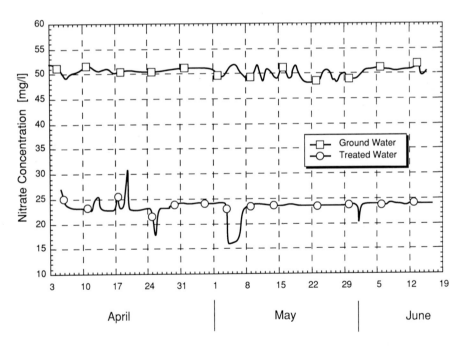

Fig. 3. Experience from the pilot studies at Münchenbuchsee, Switzerland, 1989

An important factor in any such treatment technology is the question of how much chemical dosing is required. The NITREM® process does not require any chemicals or additives, with the small exception of the regeneration salt for the electrode rinse softener (this stream usually comprises less than 1% of the drinking-water plant capacity). Thus, this process does not solve one problem by shifting the burden elsewhere. Other advantages are lower equipment costs and smaller plant sizes, and in addition, the small but significant water softening which could lower household costs and reduce detergent use.

Table 1. Typical water compositions during pilot studies at Münchenbuchsee

	Raw Water	Treated Water
Temperature [°C]	9.5	10.5
pH	7.30	7.20
Calcium [mg/L]	132	106
Magnesium [mg/L]	19.3	16
Sodium [mg/L]	6.9	6.5
Potassium [mg/L]	2.9	2.4
Chloride [mg/L]	17.6	11.0
Nitrate [mg/L]	>50	<25
Sulfate [mg/L]	22	21
Carbonate [mg/L]	467	410

An appropriate discharge and/or treatment plan for the concentrate effluent also must be considered. One possibility is mixing plant and household wastewater prior to treatment by the sewage facility. However, should a reduction of nitrates in the waste stream be desired or required, existing biological technology can be incorporated into the wastewater stream. Since the waste stream produced by the electrodialysis treatment stage is essentially drinking water with very high nitrate concentrations, a biological treatment would operate at significantly higher removal rates than a similar treatment of the drinking water itself. This leads to a plant of considerably smaller volume. This combined NITREM® bioreactor process removes nitrate from the global water household just as effectively as a biological drinking-water treatment without producing any dangers for the drinking water itself.

A deciding factor in the choice of nitrate-removal processes will always be the specific production costs for the treated water. Table 2 shows the specific investment costs for a number of processes (Rohmann et al., 1985) and for the selective electrodialysis process described in this paper. Investment costs are amortized over 15 years at 8.5% and running costs added in order to calculate the specific treatment costs.

which contribute nitrate ions to ground water. For example, the nearest well to Manda Creek (K1) was most affected by COD, turbidity and nitrate ions; this was as expected. However, the second nearest well to the Creek (K3) was less polluted than the furthest well (R). This may indicate that surface water is not the only pollution source in the area, but agricultural activities and cesspools also affect the quality of the ground water. In fact, well R, which had been selected as the reference well and had been assumed to be unpolluted, was found to contain 120 mg/L of nitrate (the highest concentration in the area). Another important observation is that although Melez and Manda Creeks are heavily polluted in respect to many quality indicators, the nitrate content is not as high as expected. This may be due to denitrification; both creeks have anaerobic conditions which can be easily detected by their unpleasant odors and bubbling gases.

Although the nitrate increase in ground water is a very important problem, there are many promising techniques for nitrate removal. For example, a study conducted by Türkman (1988) found ion-exchange techniques to be very efficient in removing nitrate.

Efficiencies for different types of resins varied between 13 and 96%. Best removal was achieved using amberlite resin (strong base type, ammonium function, chloride type). For varying nitrate concentrations, an average removal efficiency of 85% is obtained with this resin. Other average efficiencies were 77% for zerolit and 52% for amberlite weak-base type. Actually, very high removal efficiencies are not required since some amount of nitrate in drinking water is allowable without any proven adverse health effect. Thus, even less-efficient resins may be used as long as treated water contains less than 50 mg/L of nitrate at most.

References

Ağacık G (1981) Porsuk Çayının Kütahya Azot Fabrikası Atıklarıyla Kirlenmesi, Su ve Toprak Kaynaklarının Geliştirilmesi Konferansı, DSİ Genel Müdürlüğü, Ankara
Deutsch M (1963) Groundwater Contamination and Legal Controls in Michigan, Geological Survey Water Supply Paper 1691, U.S. Government Printing Office, Washington
McCracken RA, Nickerson HD (1964) Groundwater contamination in two New England communities. Public Works (February)
Prösl KH, Rimawi O (1986) Nitrates in groundwater. Bulletin of the Water Research and Study Center, University of Jordan, Amman, Jordan
Standard Methods (1975) Standard Methods for the Examination of Water and Wastewater, 14th edn.
TSE (1965) Türk Standartları Enstitüsü, İçme Suları TSE 266, Ankara

Türkman A (1981) Yeraltısuyunda Çözünmüş Maddelerin Kimyasal İlgilerinin Belirlenmesi ve Kirlenme Sorununun İncelenmesi, E.Ü. İnşaat Fakültesi Çevre Müh. Bölümü, İzmir

Türkman A (1988) Application of ion exchange techniques for nitrate removal from drinking water. Environment '88. Proceedings of the 4th Environmental Science and Technology Conference

World Health Organization (1961) European Standards for Drinking Water, Geneva

Yahşi R (1981) Su ve Toprak Kaynaklarının Kirlenmesi ve Su Ürünleri Genel Müdürlüğünün Su Kirliliği ile İlgili Çalışmaranı Su ve Toprak Kaynaklarının Geliştirilmesi Konferansı, DSİ Genel Müdürlüğü, Ankara

MITIGATING NONPOINT-SOURCE NITRATE POLLUTION BY RIPARIAN-ZONE DENITRIFICATION

L.A. Schipper,[1] A.B. Cooper,[2] and W.J. Dyck
Forest Research Institute
Private Bag
Rotorua, New Zealand

Abstract

The conversion of nitrate to nitrogen gas by denitrification has been shown to be highly active in riparian-zone organic soils. Nonpoint-source nitrate pollution of streams and lakes can be reduced by utilizing the renovative capacity of these riparian zones. The factors controlling denitrification in riparian zones are reviewed, focusing on the role of carbon supply. Results from a number of recent New Zealand studies are used to illustrate the magnitude of the riparian-zone denitrification process and the factors controlling its rate.

When comparing the relative capacity for nitrate renovation between different riparian and wetland soils it was evident that the potential rate of denitrification was limited by the concentration of available carbon. However, *in situ* denitrification rates at specific sites were shown to be limited by the concentration of nitrate, suggesting that excess untapped nitrate renovation capacity existed at these sites. At some sites, the removal of nitrate from ground water by denitrification in riparian zones was as high as 98%. By understanding the factors controlling *in situ* denitrification rates, the possibility exists for enhancing nitrate renovation by appropriate management of riparian zones.

1 Introduction

Nitrate pollution has become of increasing concern in many countries, including New Zealand. Concentrations of nitrate in ground water can exceed the World Health Organization (WHO) limit of 10 g N m^{-3} for drinking water (Burden, 1982) and eutrophication of several nationally important lakes has been attributed to increased N inputs (Viner and White, 1987).

---

[1] Schipper is also associated with the University of Waikato, Private Bag, Hamilton, NZ.
[2] Water Quality Centre, DSIR, P.O. Box 11 115, Hamilton, NZ.

NATO ASI Series, Vol. G 30
Nitrate Contamination
Edited by I. Bogárdi and R. D. Kuzelka
© Springer-Verlag Berlin Heidelberg 1991

In considering practical options for the control of nitrate pollution, New Zealand researchers have concluded that there is little that can be done by way of source control without a drastic, and economically unacceptable, reduction in farm productivity. For this reason, attention has focused on the ability of riparian zones[3] to remove nitrate from drainage water before it enters surface waters. There is circumstantial evidence that nitrate removal by bacterial denitrification may be particularly active in the riparian-zone soils of New Zealand pastoral watersheds. Nitrate concentrations in streamwaters (Wilcock, 1986) are typically an order of magnitude less than nitrate concentrations reported in the ground water which supplies the flow (Burden, 1982). In addition to the potential of riparian zones to reduce nitrate pollution from pastures, their capacity to fulfill a similar role downslope of waste treatment sites is now becoming recognized in New Zealand (Cooper, 1987; Schipper et al., 1989). A schematic diagram of N removal by riparian soils is given in Fig. 1.

The denitrification rate is influenced by a number of diverse environmental parameters (see reviews by Knowles, 1982; Payne, 1983; Fillery, 1983; Tiedje, 1988). While many of these parameters have been investigated in laboratory studies, their effects at a landscape scale are poorly understood. Factors controlling denitrification can be broadly divided into either proximate or distal regulators (Tiedje, 1988). Proximate regulators are those that affect the immediate environment of the bacterial cell, such as temperature, Ph, oxygen, nitrate and carbon concentrations. Distal regulators control the proximate regulators on a landscape scale, and include such factors as rainfall, plant growth, soil structure, and input of litter. Often a single distal factor influences more than one proximate factor; for example, moisture content can affect nitrate, carbon and oxygen diffusion. Recently, research has been directed towards identifying site or distal regulators which predict *in situ* denitrification rate at a landscape scale (e.g., Groffman and Tiedje, 1989; Cooper, in press).

In this paper we discuss the potential use of riparian-zone denitrification to mitigate nitrate pollution of surface waters. We briefly review the factors which control the denitrification

---

[3] The term "riparian zone" is often loosely used to describe a variety of margins separating water-unsaturated soils from streams, rivers, lakes, and wetlands. Riparian zone soils are often high in clay and organic matter, and are saturated by laterally moving ground water which fluctuates in height (Lowrance et al., 1985).

rate and use examples from our recent field studies to illustrate the consequences of some of these factors at the riparian-zone scale.

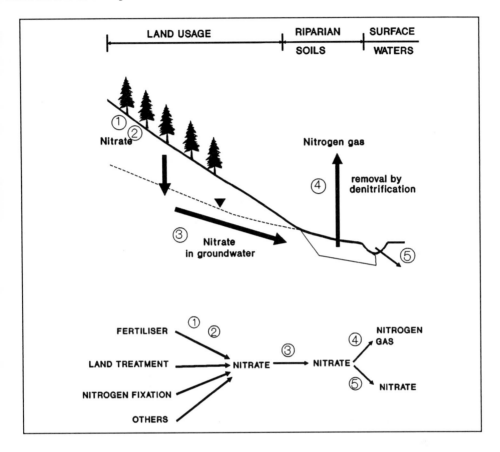

Fig. 1. A schematic diagram of the pathway and processes of nitrogen, including (1) mineralization, (2) nitrification, (3) ground-water leaching, (4) denitrification, and (5) leaching to surface waters

2 Factors Regulating Denitrification in Riparian Zones

2.1 Oxygen

Denitrification generally only occurs in sites which are not aerobic (Sextone et al., 1985), as the presence of oxygen causes a reversible inhibition of the bacterial enzymes involved in the denitrification process (Ferguson, 1987). Any soil characteristic that influences either the diffusion or aerobic consumption of oxygen will directly affect the aerobic status of the soil and, therefore, the rate of denitrification. For example, in agricultural soils, increased soil-water content decreases oxygen diffusion and so may increase the denitrification rate (Parkin and

Table 1. Denitrifying enzyme activity (DEA) from a range of ecosystems

Ecosystem	DEA (ng N g^{-1} h^{-1})	Reference
Forest soils	84 - 392	Tiedje, 1988
Agricultural soils	5.6 - 280	Tiedje, 1988
Eutrophic sediment	6,440	Tiedje, 1988
Forest riparian soils	170 - 3750	unpubl. data by authors
pasture riparian zones		
(i) organic soils	2610	Cooper, in press
(ii) mineral soils	44	Cooper, in press

2.5 Nitrate Supply

Nitrate is used as the electron acceptor during the denitrification process, and hence nitrate concentration has a strong influence on the denitrification rate. The availability of nitrate to denitrifying bacteria is dependent upon the rate of production of nitrate (nitrification); the rate of nitrate consumption by nondenitrifiers, including plants and bacteria, and the rate of nitrate diffusion through the soil (Tiedje, 1988). In most soils, the production of nitrate by nitrification is a necessary precursor for denitrification. However, riparian organic soils are usually anaerobic and little nitrification can occur; hence denitrification is dependent upon an influx of nitrate from upslope ecosystems.

The dependence of denitrification on nitrate has been examined in the laboratory but not well investigated at the landscape scale. In laboratory studies, Murray et al. (1989) showed that denitrifying bacteria in agricultural soils and pond sediments had a very high affinity for nitrate and maintained this affinity even in environments that received high nitrate inputs. Reddy et al. (1982) examined denitrification kinetics in a number of organic and mineral soils and showed that denitrification kinetics were controlled by both nitrate and available-carbon concentrations.

We investigated the regulatory role of nitrate in riparian soil denitrification and examined a riparian zone below a sewage land-treatment system (Tairua Forest, New Zealand). This riparian zone had previously been shown to remove 98% of incoming nitrate and it was inferred that denitrification was the mechanism responsible (Schipper et al., 1989). Thirty-five soil cores were taken from a grid sampling in a 2 m by 2 m section of the riparian zone and measurements made of on-site denitrification, WSC, and nitrate concentration (methods used were similar to those described in Cooper (in press)). All soil cores were saturated and it was found that the denitrification rate and nitrate concentration decreased down the riparian zone (Fig. 2), while the WSC concentration (not shown) remained constant. This implied that the denitrification rate within the riparian zone was limited by the nitrate concentration rather than by available carbon. Similar conclusions were reached for denitrification within the riparian organic soils of a pastural catchment (Cooper, in press).

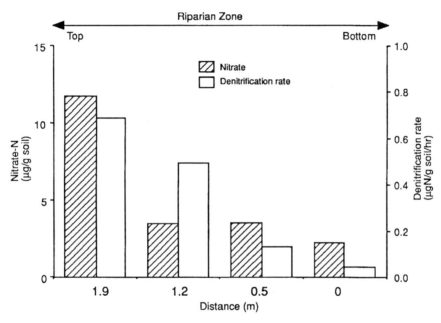

Fig. 2. Nitrate concentration and on-site denitrification rate measured from soil cores within an organic riparian zone (Tairua Forest, NZ)

2.6 Influence of Hydrology

For denitrification to be an efficient process in riparian zones requires not only suitable conditions for the expression of denitrifier activity but also sufficient contact between nitrate-laden ground waters and active organic soils. This hydrological control on the overall efficiency of the denitrification process has been demonstrated in several studies. Warwick and Hill (1988) found that little nitrate removal occurred in a riparian zone of a small some nitrate bypassed treatment by flowing across the surface of the riparian zone. Still, this removal rate is very high compared to other ecosystems (Table 2). The difference in riparian-zone denitrification rates between our study and that reported by Cooper (in press) cannot be explained by differences in the potential denitrifying enzyme activities (Table 1), and probably reflects differences in nitrate-supply rates. This further suggests that in some cases riparian-zone denitrification is limited by incoming nitrate.

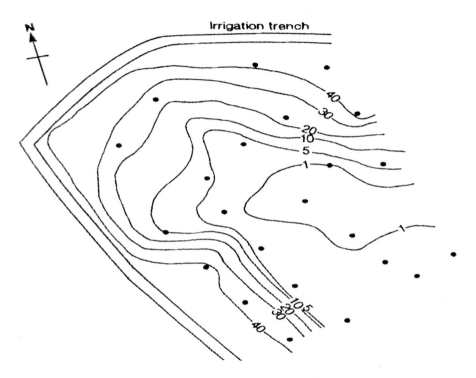

Fig. 5. Isopleth plot of plateau NO_3^--N/Cl^- ratio ($\times 10^{-3}$) at the gully head. Before irrigation the average ratio was 2.2×10^{-3}. Solid circles are well locations

Table 2. Landscape rates of denitrification

Site	Denitrification kg N/ha.yr	Reference
Forest soils	0 - 40	Groffman and Tiedje, 1989
Grassland	8 - 14	Bijay-Singh et al., 1989
Arable land	50 - 87	Bijay-Singh et al., 1989
Riparian soils (no added nitrate)	26 - 183	Cooper, in press
Riparian soils (added nitrate)	700	unpubl. data by authors

4 Management and Recommendations

Management techniques for optimizing N renovation within riparian zones have not been extensively explored. However, from our experience it is possible to make some intuitive recommendations. First, despite the high organic-matter content of many riparian soils it is likely that sustained denitrification is still limited by the supply of carbon. Addition of mulched plant biomass to artificial wetlands increased N renovation (Gersberg et al., 1984), but the use of a similar approach to enhance denitrification in riparian zones may not be economically feasible. An alternative option for increasing carbon supply to denitrifiers is to establish highly productive and readily decomposable vegetation within riparian zones which would increase carbon input and allow an organic soil to develop. Secondly, nitrate-renovation efficiency on the landscape scale is likely to be a function of the extent of contact between active riparian zones and the nitrate-laden ground water moving towards surface waters. Management practices which decrease this contact, such as land drainage and channelization, should be avoided. Management practices which increase this contact, such as the siting of land-treatment schemes upslope of riparian organic soils and wetlands, should be encouraged. Although the potential exists for optimizing N renovation within riparian zones, further research is required to realize this potential.

We recommend research in the following areas:

1. The effect of organic-matter quality on denitrification in riparian soils.
2. The relationship between productivity and decomposability of various riparian plants and carbon supply for denitrifiers.
3. The effect of water residence times in riparian soils on nitrate-renovation efficiency.
4. The effects of increased nutrients and water on the long-term renovation capacity and ecology of riparian zones. It should be recognized that these zones often serve many valuable ecological functions (e.g., flora and fauna habitat), and changes to these functions in response to upslope management need to be understood.

Acknowledgements

The authors thank Christine Mees and Phil Squire for their invaluable comments on this manuscript. Chris Harfoot, Warrick Silvester, Peter Stevens, and Paul MacFarlane all form part of L.A. Schipper's doctorate supervisory committee and provided invaluable instruction compounds from nitrates in drinking water have become issues of increasing concern (Shuval and Gruener, 1977; NAS, 1978; White, 1983). The contribution of nitrate to birth defects has been studied by Dorsch et al. (1984), who found a four-fold increase in risk of congenital malformations for pregnant women drinking water containing greater than 10 mg/L as nitrate-nitrogen.

A major contributor of nitrate is the on-site disposal of wastewater (Minear and Patterson, 1973; Canter and Knox, 1985). While the conventional septic-tank soil-absorption system provides for Biological Oxygen Demand (BOD) and Total Suspended Solids (TSS) reductions, nitrogenous compounds are only nitrified, thereby contaminating ground water with nitrate. In a recent study of septic-tank systems in Maine, Black and Struchtemeyer (1982) found nitrate concentrations exceeded the 10 mg/L as nitrate-nitrogen standard in 71% of the systems monitored. In neighboring New Brunswick, monitoring of 300 wells revealed 16% above the Canadian drinking-water standard of 10 mg/L as nitrate-nitrogen (Brennan, 1984). In New Jersey, the Pineland Commission has set a standard of 2 mg/L as nitrate-nitrogen in the ground water at the boundary of each house lot to prevent further deterioration of ground water from septic-tank systems (Ehrenfeld, 1984).

Evidence of denitrification in conventional soil-absorption fields has been reported by Reneau (1977, 1979), who found that zones of denitrification (anaerobic conditions) were created by a fluctuating water table. In contrast, Viraraghavan and Warnock (1976) and Cogger and Carlile (1984) did not observe denitrification in the conventional and alternative on-site systems studied. Reasons cited include a high water table, cold temperatures, lack of a suitable energy source and/or effluent nitrogen not being nitrified. In laboratory columns used to simulate mound systems, Magdoff et al. (1974) found that a low organic-carbon supply limited denitrification, and nitrogen removal was only 32%. However, Harkin et al. (1979) observed an average reduction of 44% in operating mound systems. In some instances they observed higher values in dosed systems.

Several researchers have developed modified systems to improve nitrogen removal for on-site systems. Andreoli et al. (1979) used methanol as the basis for their modified septic-tank soil-absorption system. An impermeable shield was located below the distribution pipe to collect the nitrified effluent and direct it to an impermeable pan to which methanol was added. The overall removal rate was only 36% since much of the nitrified effluent missed the pan because the horizontal hydraulic conductivity of the soil was greater than the vertical conductivity. Even if this design flaw could have been overcome, the system required a feed system to provide the external organic carbon (methanol), system construction was complex, and the system was over two meters deep.

Laak et al. (1981) developed a modified system (RUCK system) which is capable of achieving 80% nitrogen reduction by separating the greywater from the blackwater. By using greywater instead of methanol, the researchers effectively circumvented the problems associated with an external carbon source. However, two septic tanks, a water-tight sand filter, and a seepage bed are necessary for successful operation. In addition, household plumbing must be designed to provide the necessary separation of the wastewaters. Finally, the design is not conducive to retrofitting existing systems nor is it easily integrated into current practice.

A similar system (anoxic gravel filter, two sand filters, and a pump) developed by Piluk and Hao (1989) achieved good nitrogen removal without requiring separation of the greywater and blackwater. By maintaining a recirculation ratio of 3.3 they achieved a 70% reduction in total nitrogen. Like the RUCK system, this system requires extensive construction and, in addition, mechanical pumping.

Research by Niimi (1981) suggested that septic-tank effluent could be used as the carbon source without the extensive facilities required by Laak and Piluk. In a review of the Niimi Process, Yahata (1981) claimed that both reduction and oxidation took place in the same area; that is, denitrification under reducing conditions occurred in the same location in which nitrate oxidation previously occurred, an impermeable pan filled with "capillary sand" located 10-15 cm below the wastewater distribution pipe. The natural, intermittent flow pattern resulted in anaerobic bacteria working when the wastewater arrived, while aerobic bacteria were active when the wastewater stopped flowing (Yahata, 1981). Limited supporting data were presented, but the total nitrogen reduction ranged from only 37 to 47%. Further evaluation by Dillaha et al. (1985) showed limited success with the Niimi process, primarily due to the influence of soil types on unsaturated flow, nitrogen dynamics, hydraulic loading, and waste transport. They reported only 27 to 38% nitrogen removal, which was not statistically different from the conventional trench tested. The Niimi System has also been tested by Reed et al. (1989), who reported good organic removal but did not look at nitrogen removal.

There is potential for denitrification in on-site systems; in fact, good nitrogen removal has been documented even in some existing on-site systems. In these isolated cases, natural conditions have accidently combined to provide the unique environment required by nitrifying and denitrifying bacteria. Yet the only designed systems giving good nitrogen removal are relatively complex and expensive. The availability of an effective, low-technology denitrification unit remains to be demonstrated.

The principal objective of this project was to develop a passive denitrification technology to remove nitrate from septic-tank effluent. In order to be effective in on-site applications, such a system must offer several advantages over existing technologies:

- mechanically simple and low-maintenance requirements
- no auxiliary organic carbon feed
- no pumping or recirculation requirements
- no extra components
- minimal changes in current technology

On the basis of these objectives, a laboratory study was conducted of two basic designs. The study addressed the following questions: (1) could the work of Niimi be improved upon to achieve an acceptable reduction in nitrate-nitrogen? (2) could a completely artificial system be developed to replace the conventional soil-absorption system? Researchers hypothesized that a plastic trench segment incorporating the distribution pipe, filtering media, and an impermeable tray layer into a self-contained unit would provide the required environmental conditions for nitrification and denitrification within the same area.

The latter system, if successful, could be installed just as easily as distribution pipes are in current construction practice. Retrofitting existing soil-absorption trenches would also be easily accomplished. No additional facilities or metering systems would be needed. The only function of the natural soil would be the assimilation of the treated water. In those cases where the native soil was unsuitable, the treated water could easily be collected and discharged to surface water as is the current practice with sand filter effluent.

2 Methodology

2.1 Experimental Reactor Designs

The project was divided into two major elements: (1) system development using sand for filtration and (2) completely synthetic construction of the filtration system. In Part 1, the rectangular tanks (28 cm × 28 cm × 89 cm) contained 86 cm of sand (coefficient of uniformity=6) with the bottom portion under saturated conditions (anaerobic zone for denitrification). The saturated zone was maintained by raising the water level as shown in Fig. 1. A second, separate tank contained sufficient sand to mimic only the aerobic zone of the experimental reactor; the results from this tank allowed quantification of nitrification. To ensure uniform distribution, a layer of geotextile material (Exxon Tiger Drain) was placed on the top of the column. This geotextile consisted of two bonded materials, a corrugated (egg-carton-like) open plastic sheet with an attached cloth of loose weave fiberglass, for a total thickness of approximately 2 cm. Two more layers of geotextile were placed at the bottom of the tank to prevent clogging of the outlet. Septic-tank effluent collected from a nearby on-site system (Table 1) was applied at loading rates ranging from 2.1 to 4.2 cm/day, depending upon the experimental design.

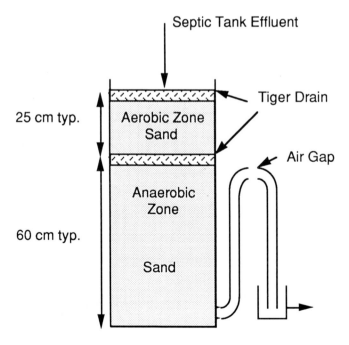

Fig. 1. Schematic of nitrification and denitrification reactor

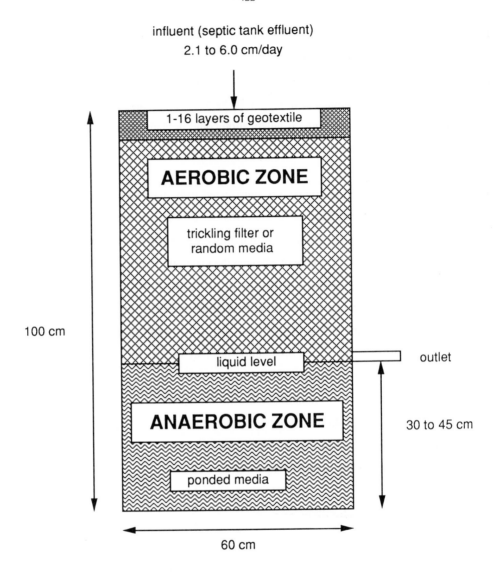

Fig. 3. Generalized artificial trench unit design

were used to measure the COD. Since the COD of the effluent was very low, low-range vials (0-150 mg/L) were used in this study.

Ammonia- and Organic-Nitrogen were measured by the macro Kjeldahl method in accordance with Standard Methods (APHA, 1985). The ammonia-nitrogen was also measured using the ion-specific electrode technique (ammonia electrode - Orion Model No. 95-12).

Nitrate-Nitrogen was measured using the ion-specific electrode (ammonia electrode- Orion, Model No. 95-12). The meter was calibrated using standard solutions (100 ppm nitrate and ammonia). Initially, ammonia-nitrogen present in the sample was measured, then the nitrate-nitrogen present in the sample was reduced to ammonia using a reducing agent (titanous chloride), and the resultant reading gave the total ammonia-nitrogen concentration. The difference between the two readings was the amount of nitrate-nitrogen in the sample.

Soluble Organic Carbon. The samples were filtered to obtain the soluble organic carbon. The filtered samples were oxidized by the wet oxidation method using potassium persulfate. Carbon dioxide was then measured using an infrared carbon analyzer (O.I. Corporation, Model No. 524C).

3 Results and Discussion

In order for denitrification to be effective, it is necessary to have an organic carbon source and an anaerobic environment, conditions conflicting with the oxidizing aerobic requirement for nitrification. Thus, denitrification must follow the nitrification step in the soil-absorption bed. Unfortunately, as the septic-tank effluent is being nitrified in the soil-absorption trench, its organic carbon is also being oxidized, thereby reducing the efficacy of the septic-tank effluent to serve as an organic carbon source.

Consequently, various depths were considered to establish the relative depths of the aerobic and anaerobic zones such that the septic-tank effluent would be nitrified prior to entering the impermeable pan, but sufficient organic carbon would remain for denitrification. According to Walker et al. (1973), the nitrification step does not require a great distance in soil, and they reported ammonia levels decreased rapidly in the top 10 cm of soil beneath the drainfield. In the Niimi process the total depth of treatment was only 20 cm of sand, again with 10 cm devoted to nitrification (Dillaha et al., 1985). The remaining 10 cm represented the anaerobic or denitrification zone. Unfortunately, Dillaha et al. (1985) did not find very good nitrogen removal, whereas Warnock and Biswas (1982) found effective denitrification with about 25 cm of sand.

In Part I, preliminary testing indicated that 55 cm gave excellent nitrification, but the remaining 30 cm were insufficient for denitrification. Thus in the majority of testing the aerobic zone was represented by 25 cm of sand and the denitrification zone by 60 cm. The aerobic zone was found capable of removing greater than 97% of the BOD, and the removal

advantages of using septic-tank effluent are that there is no additional cost, and it is easily available and relatively nontoxic.

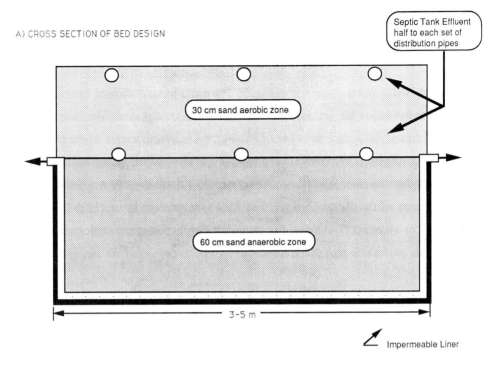

A) CROSS SECTION OF BED DESIGN

Septic Tank Effluent half to each set of distribution pipes

30 cm sand aerobic zone

60 cm sand anaerobic zone

3–5 m

Impermeable Liner

B) CROSS SECTION OF TRENCH DESIGN

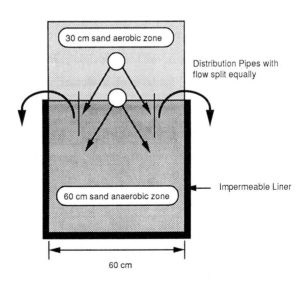

30 cm sand aerobic zone

Distribution Pipes with flow split equally

Impermeable Liner

60 cm sand anaerobic zone

60 cm

Fig. 5. Proposed cross sections of NRID on-site systems to remove nitrogen

Unfortunately, it was found that trickling filter and random media were not appropriate for the application of this research. This is primarily, it is believed, because of the inadequate surface area to volume ratios, which resulted in retention times that were too low for adequate treatment of the septic-tank effluent. It may be that plastic media could be utilized successfully if much more densely packed configurations could be used. Nevertheless, a simple system using no moving parts or external source of organic carbon has been developed that is capable of meeting a nitrate-nitrogen limit of 10 mg/L.

In comparison to the other systems used for nitrogen removal, the NRID System offers four advantages:

1. No need for an external carbon source
2. Mechanically simple
3. Easy to adapt to the current construction practice
4. Reduction of nitrogen without adding other pollutants

The system operated for six months with no signs of clogging of the sand media. The life expectancy of the system seems to be promising, but long-term conditions have to be evaluated.

Acknowledgements

The activities on which this publication is based were financed in part by the Department of Civil Engineering, University of Maine and the Department of the Interior, U.S. Geological Survey, through the University of Maine Environmental Studies Center. The contents of this publication do not necessarily reflect the views and policies of the Department of the Interior, nor does mention of trade names or commercial products constitute their endorsement by the United States Government. The assistance of Ann K. Lounsbury is gratefully acknowledged.

References

Alexander M (1977) Introduction to Soil Microbiology. John Wiley & Sons, Inc. New York
American Public Health Association (1985) Standard Methods for the Examination of Water and Wastewater, 16th edn., APHA, New York
Andredakis AD (1985) Organic matter and nitrogen removal by an on-site sewage treatment and disposal system. Wat Res 21(5):559-565
Andreoli A, Reynolds R, Bartilucci N, Furgione R (1979) Nitrogen removal in a subsurface disposal system. Prog Wat Tech 12:967-976
Black RW, Struchtemeyer RA (1982) Effectiveness of septic systems in safeguarding the groundwater in Maine. OWRT Completion Report Proj. A-052-ME. Land and Water

CONTROL OF NITROGEN SOURCES AND PRINCIPLES OF TREATMENT

A. Samsunlu
Department of Environmental Engineering
Istanbul Technical University, 80626 Maslak
Istanbul, Turkey

Abstract

Nitrogen-removal efficiencies were calculated in relation to sludge ages. The effluent of a physical-treatment plant was passed through a plant-scale biological unit where nitrification and denitrification occurred. A biological-treatment unit consisting of five cascades was used as a pilot-scale plant, and the respirometric activity and nitrogen concentration in different forms were measured. The variation of ammonium concentrations within the reactor under various recycle ratios and sludge load conditions is presented.

1 Introduction

Especially in industrialized countries, it is no longer possible to use polluted surface waters for drinking-water supplies. Nitrogen pollution in surface waters originates mainly from domestic and industrial wastewater, and drainage and surface runoff from agricultural areas. The amount of nitrogen in domestic wastewater discharges is in the range of 8 to 15 gr/day/ person (Helmer and Sekoulov, 1979).

High nitrogen loads in receiving waters create several problems (Diesterweg, 1985):

1. The greater part of ammonia nitrogen (NH_4-N) in river waters readily converts to nitrate. The oxygen required for this process is drawn from the water; when river oxygen levels fall below 4 mg/L fish stocks are threatened. This process occurs more rapidly and has more serious consequences as water temperatures rise.

2. At pH values over 7, the ratio of ammonium (NH_4^+) to free ammonia (NH_3) shifts in favor of the free ammonia, which is toxic to fish and other aquatic organisms. The toxicity limit of ammonia is in the range of 0.2 to 2 mg/L for fish, and 0.2 to 9 mg/L for other aquatic animals.

3. Increased levels of NH_4^+ and NO_3^- in stagnant waters promote plant growth and can lead to eutrophication.

NATO ASI Series, Vol. G 30
Nitrate Contamination
Edited by I. Bogárdi and R. D. Kuzelka
© Springer-Verlag Berlin Heidelberg 1991

4. Drinking-water quality is threatened by high nitrate and nitrite concentrations, which can lead to infantile cyanosis.

The various forms of nitrogen that are present in nature and the pathway between them are shown schematically in Fig. 1. Table 1 shows concentrations of nitrogen compounds in domestic and some industrial wastewaters.

Table 1. Concentrations of nitrogen compounds in domestic and some industrial wastewaters

Source	Type of Nitrogen	Concentration (mg/L)		
		Strong	Medium	Weak
Domestic				
	Nitrogen (total as N)	85	40	20
	Organic Nitrogen	35	15	8
Constituent	Free Ammonia	50	25	12
	Nitrite	0	0	0
	Nitrate	0	0	0
Industrial				
Adhesives and sealant	Ammonia	20		5
Explosives manufacturing	Ammonia	125		45
	Nitrate	7000		0.4
Iron and Steel	Ammonia	7		3.9
Leather tanning and finishing	Ammonia	680		0.4
Organic chemicals	Ammonia	50		1
Petroleum refining	Ammonia	168		1
Timber products	Ammonia	80		20

2 Nitrogen Removal in Wastewater

2.1 Treatment Stages

Generally, domestic wastewater is treated in a two-stage (mechanical and biological) process in which suspended solids and dissolved organic matter are removed. Advanced treatment methods are then used to remove the untreated or slightly treated pollutants. Fig. 2 shows different treatment alternatives (joint and separate) for industrial and domestic wastewaters. Biological, physical and chemical processes can be used to remove nitrogen, ammonia nitrogen and nitrate nitrogen at various points in the treatment process.

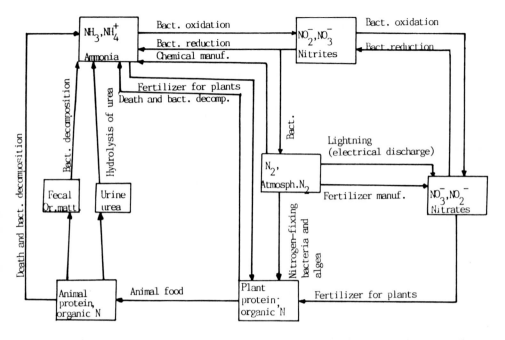

Fig. 1. Nitrogen cycle (after Young et al., 1975)

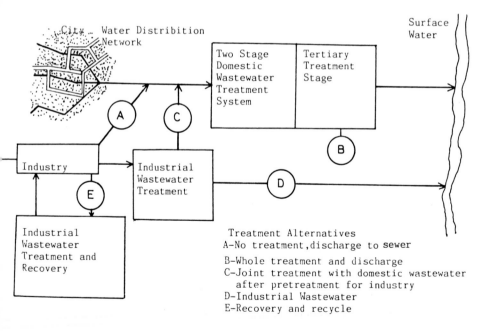

Fig. 2. Treatment alternatives for industrial and domestic wastewater (after U.S. EPA, 1973)

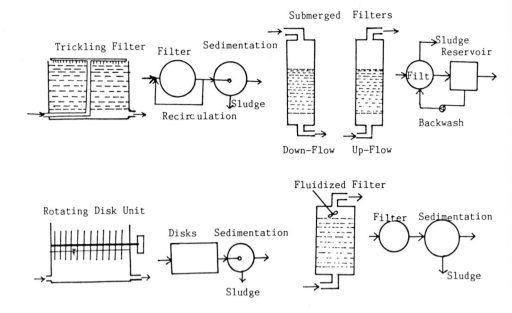

Fig. 5. Biofilm reactors employed in wastewater treatment

3.1 Description of the Model

Raw sewage from the mechanical-treatment process was introduced to a biological-treatment unit (Fig. 6) consisting of five cascades. The first part was stirred, and the subsequent four cascades were aerated with diffused air. The aeration pipes are placed at the two sides of the tank. Apart from the biological unit there was also a primary sedimentation tank and equipment and pumps for recycling. The first cascade was designed for the purpose of denitrification, while the others were designed with the aim of nitrification. In order to achieve a higher nitrogen removal rate, the recycle ratio, Q_R, and the influent water flow rate, Q_g were changed, and NH_4-N, NO_3-N, NO_2-N measurements were performed in each cascade in the biological unit for every case. The oxygen level in the water was maintained at a constant 1.5 mg O_2/L, but some variation in the dissolved oxygen concentration occurred, and the DOC level occasionally fell below 1 mg/L.

In the study, the respirometric activity and nitrogen concentrations in different forms were measured, including NH_4-N, total nitrogen, NO_3-N, and NO_2-N. Concentrations of the different forms of nitrogen were determined by standard methods (APHA, 1985), and pH, alkalinity, COD and BOD_5 were measured. The measurements were performed on the influent (precipitated raw sewage), every cascade of the biological unit (five cascades), and the effluent water.

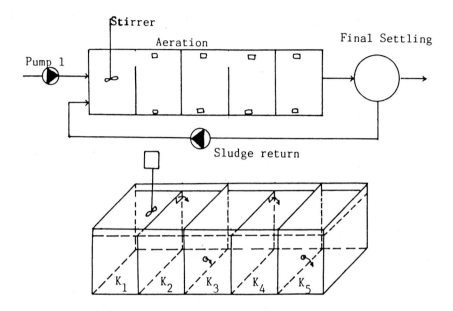

Fig. 6. Biological nitrogen-removal unit model

3.2 Experimental Results

In the experiment, samples were taken from inlets and outlets of numbered cascades (from one to five). Alkalinity, suspended solids, COD, BOD_5, total nitrogen, ammonia nitrogen, nitrate nitrogen, nitrite nitrogen, total phosphorus, temperature, volatile suspended solids, dissolved oxygen, sludge volume index and respirometric activity tests were performed on the samples. The inlet flow rate is expressed as Q_g and the recycled rate as Q_R. Influent flow rates, recycled flow rates, and variations of NH_4-N in influent and recycle flow rates as maximum and average values are provided in Table 2.

In Fig. 7, the NH_4-N values versus different recycle percentages are plotted. NH_4-N concentrations in the effluent must be under 3 mg/L. Since this limitation is only met by recycle percentage of 600, it is possible to get a value under the limit in the second cascade. The number of the cascades can thus be reduced.

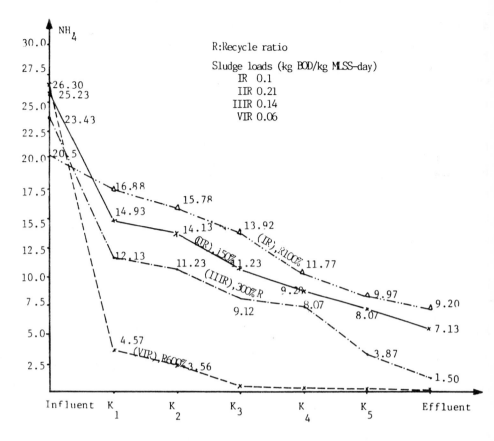

Fig. 7. Variation of ammonium concentrations within the reactor under various recycle-ratio and sludge-load conditions

Table 2. Performance data for various sludge loadings

	IR	IIR	IIIR	VIR
Q_g (1/min)	0.25	0.750	0.500	0.125
Sludge Load (kg/kg-day)	0.08-0.12	0.15-0.28	0.10-0.20	0.03-0.08
Average Sludge Load	0.1	0.21	0.14	0.06
Q_R 1/min	0.750	0.750	0.750	0.750
Recycle Ratio (%)	300	100	150	600
NH_4-N Ranges Monitored at Effluent Water (mg/L)	2- 2.5	2.5- 19.8	7.8- 11.5	Not meas.
Average Concentrations (mg/L)	1.5	9.20	7.13	Not meas.

The sludge load in the treatment plant can be 0.3 kg/kg-day and higher if carbon removal only is considered. However, if nitrification is also required, the sludge load must be approximately 0.1 kg/kg-day or lower (Samsunlu, 1986).

4 Conclusions

Nitrogen compounds play an important role in the ecological balance. Nitrogen is discharged every day in increasing amounts from domestic- and industrial-wastewater sources. If nitrogen is not controlled at the wastewater source, eutrophication can occur in surface waters.

Nitrogen removal rates in conventional-treatment systems are very low and nitrogen removal from wastewaters is very expensive. To remove nitrogen in biological-treatment systems by nitrification and denitrification processes, it is necessary to design and operate biological-treatment plants. Nitrification and denitrification can be successfully achieved by activated-sludge or submerged-filter systems. However, in order to achieve more effective nitrogen removal, it is necessary to add a tertiary physical-chemical treatment process following biological treatment.

References

APHA, AWWA, WPCF, (1985) Standard Methods for the Examination of Water and Wastewater, 16th Edition

Diesterweg G (1985) Tower-Biology and Its Applications for the Nitrification/Denitrification of Ammonia-Rich Wastewater. Proceedings of the 40th Industrial Waste Conference, Purdue University, Purdue, Illinois

Harremoes P, Gönenç ijE (1983) The Applicability of Biofilm Kinetics to Rotating Biological Contactors, Preprint for EWPCA/IAWPR International Seminar on Rotating Biological Discs. Stuttgart

Helmer R, Sekoulov I (1979) Weitergehende Abwasserreinigung Deutscher Fachschriften Verlag. Mainz-Weisbaden

Samsunlu A (1986) Die Uberlegungen Uber den Einsatz der Nitrification und Denitrification-Kationsstufe bei der Klaeranlagen-Plannung in der Turkei. VI Deutsch-Turkisches Seminar out den Gebiet des Technischen Unveltschutzes. Sonderheft, Stuttgart, p 65-94

U.S. EPA (1973) Physical Chemical Wastewater Treatment Plant Design. EPA Tech. Transfer Seminar Publication

U.S. EPA (1975) Process Design Manual for Nitrogen Control. Office of Technology Transfer

Young JC, Baumann ER, Wall DJ (1975) Packed Bed Reactors for Secondary Effluent BOD and Ammonia Removal. J Water Poll Control Fed 47(1)

U.S. ENVIRONMENTAL PROTECTION AGENCY PROGRAMS RELATED TO THE AGRICULTURAL USE OF NITRATES

T.L. Amsden
Acting Director
Water Management Division
U.S. Environmental Protection Agency
726 Minnesota Avenue
Kansas City, Kansas 66101 U.S.A.

Abstract

The Environmental Protection Agency (EPA) is embracing new principles and tools to meet the challenge of nitrate management. We are adopting the fundamental philosophy that less nitrate usage is better than more. In addition, we are using innovative tools, including wellhead protection as an alternative to treatment at the tap, geographic targeting, demonstration and education, and economic incentives and disincentives.

1 Introduction

As we shift increasingly toward risk assessment as a basis for environmental management in this country, agricultural chemicals are moving to the top of our priority list because of the large geographic areas affected and the large populations potentially exposed. In contrast, hazardous waste sites (Superfund sites), which have traditionally been of higher priority, now fall in priority because they affect relatively small geographic areas and populations.

As a consequence of our increased focus on agricultural chemicals, the U.S. Environmental Protection Agency (EPA), along with a wide range of other organizations at the federal, state, local, and private levels, is involved in a variety of efforts and initiatives relating to the agricultural use of nitrate.

2 Programs and Initiatives

2.1 Characterizing the Problem

A number of efforts are underway to determine the extent of ground-water contamination resulting from the agricultural use of nitrate. The National Pesticide Survey, which is our

NATO ASI Series, Vol. G 30
Nitrate Contamination
Edited by I. Bogárdi and R. D. Kuzelka
© Springer-Verlag Berlin Heidelberg 1991

conditions as well as other factors relating to the degree of contamination, labor and materials costs, and the degree of treatment desired.

During the panel discussion, there were questions from the participants as to whether nitrate treatment is a realistic method of nitrate-contamination control. There seemed to be agreement that treatment can be considered as a short-term solution whereas effective contamination control must be achieved through implementation of preventative measures at the source of contamination.

V. INTEGRATION: THE SYSTEMS APPROACH

system (Hartman, 1982). If the human is exposed to high concentrations of nitrates, the risk of cancer and noncancer can increase to unacceptable levels. The control system is responsible for administering regulatory policies, prevention neasures, and actions to protect human and ecosystem components.

The interrelationships among the system components are represented by the system couplings. Fig. 1 shows a representation of the system components and their coupling. The geophysical system receives nitrate contamination from a variety of sources, including point and distributed sources. The geophysical system provides the distribution system for the nitrate by contaminating the water supply. The human system component is exposed to nitrate contamination through ingested water. The effect of the nitrate is observed through the incidence of medical conditions such as cancer. The control system establishes a set of preventive and mitigating actions that can be used to counter the effects of nitrate pollution on human beings.

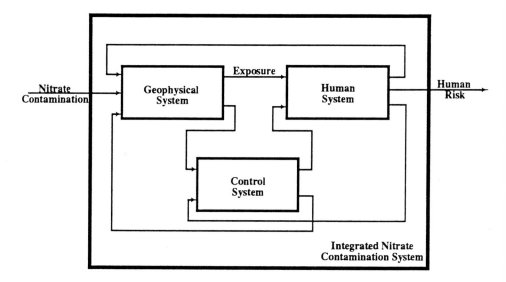

Fig. 1. Coupling between system components of the integrated nitrate-contamination system

Each of the component systems has been the focus of research by many specialists. The development of models of each of the components has been the object of numerous investigations in recent years. The papers presented at the NATO Advanced Research Workshop on Nitrate Contamination have been classified according to the scheme presented in Fig. 1. Fig. 2 shows the selected classification: many of the contributions deal directly with the system components and several others address the interrelationship between the components.

The focus here is not to study the details of each component system, but rather to establish the relationships between the components, and to present an integrated framework for the evaluation of alternative system designs. In approaching the total system as an integrated coupling of the system components, researchers in both fields can concentrate on providing compatible data to one another. To establish a framework for comparing alternative control system designs, it is necessary to determine the basis for the comparison of alternative designs. The two most important factors are cost and performance. These two factors are called figures of merit that must be traded off in selecting a design. While a specific design may have a very high degree of performance, it may have a price that makes it an undesirable alternative.

Each figure of merit, cost and performance, is composed of many individual performance indices. For example, cost consists of capital and operational costs as well as any legal costs. Performance consists of the individual and population risks of nitrate exposure, and measures such as the maximum exposure dose or length of time that exposure dose exceeds recommended levels. In addition, the figures of merit must contain information that reflects the decision maker's preference structure.

The decision maker may be an individual but is most likely a government agency that represents the interests of the population. The structure for designing alternative control systems must consider all of the important information.

In this paper each of the important factors in the design is addressed. First, a simple model structure is developed to represent the important features of the geophysical and the human components of the integrated nitrate system. The modular or component structure presented here is intended to allow specialists to utilize their detailed models of the components. However, from a systems perspective, the structure presented highlights features that are important and significant in the design of a "best" control system. On the basis of this model structure, performance measures are defined that can be used to compose figures of merit. The and for continuous time systems, the state transition function is the form

$$\frac{ds(t)}{dt} = F(s(t), x(t)) \qquad [2]$$

The state transition function provides the essential link between the elements of the system model. The outputs from the system may simply be the state itself or may be measures that reflect the behavior of the system. For example, the output of the human system may be the cancer risk of an individual or the cancer risk of the population which reflects the individual risk and the exposed-population distribution. The readout function provides a relationship between the system states and the system outputs. The time scale of the system represents the relationship between model time and physical time. It also distinguishes between discrete- and continuous-time systems.

Each of these models also contains various uncertainties in the elements discussed above. Possible sources of uncertainty include the following:

1. Natural uncertainty due to rainfall/runoff, the motion of water through the saturated and the unsaturated zones, human demand for water, diffusion processes of the chemicals, and interaction between the nitrates and other chemicals

2. Model uncertainty due to the difficulty or impossibility of representing reality with mathematical models

3. Parameter uncertainty due to inexact knowledge and finite sample data for parameter estimation, including measurement noise

4. Economic uncertainty due to the inability to forecast accurately cost factors such as construction, operation, and maintenance

5. Technological uncertainty due to difficulty in forecasting the technological advances such as low-cost, high-quality water purification techniques, modeling advances, and measurement techniques

For the purposes of the discussion presented here, the question of model uncertainty only complicates the problem; hence it will not be directly addressed. However, the above important issues should be included in a detailed study of the nitrate pollution problem. The uncertainties in the input processes and contamination from point and distributed sources are propagated through the deterministic dynamics and affect the states and outputs from which the performance indices and figures of merit are estimated. These uncertainties will have a significant effect on the control system design.

In the following three sections, system models for the three component systems are defined. The approach taken here is to capture the overall system structure without dwelling on the specific model details. The most important consideration in defining the component models is for the models to provide the information necessary for estimating the performance indices and figures of merit. Thus, it is very important that decision makers specify the most important requirements when they place demands on the system or on any system that is affected by the nitrate system. To this end, it is necessary that all requirements be defined prior to the definition of the models that will be used in the design process.

2.2 Geophysical System Model

The geophysical system is primarily responsible for the time and spatial distribution of nitrates. To represent this spatial structure the geophysical system can be decomposed into a grid or matrix of small geographical areas, which, in practice, could be part of a geographic information system. Each cell of the matrix will be modeled as a system component that is coupled to each of its neighbors (Duckstein, 1982). Fig. 3 depicts this spatial decomposition.

Consider the (i,j)th component. Let the mathematical system model be denoted

$$GEO\,(i,j) \;=\; (X(i,j), S(i,j), F(i,j), Z(i,j), G(i,j), T(i,j)) \qquad [3]$$

It will be assumed that the time scale, $T(i,j)$ is identical for every component of the geophysical system. For the purpose of the model presented here, the time scale will be assumed to be discrete with each time unit representing, for example, one day. Hence,

$$T(i,j) \;=\; 0,1,2,...$$

The input to the (i,j)th component at any point in time $t\epsilon T$ will represent the contamination of the component from each neighboring component as well as contamination from point and distributed sources and the effects of the control system. The inputs from distributed sources are affected greatly by factors such as rainfall/runoff, irrigation, and the natural topology of the terrain. All of these factors should be accounted for when modeling the input of nitrates from distributed sources. Since the structure is chosen to be general enough to accommodate many control system design alternatives, it is assumed that the control input represents the change in nitrate concentration that a specific control policy can achieve. This representation will allow for treatment of the water supply, regulation of the use of nitrate-based fertilizer, or regulation of the location of waste treatment facilities and stockyards.

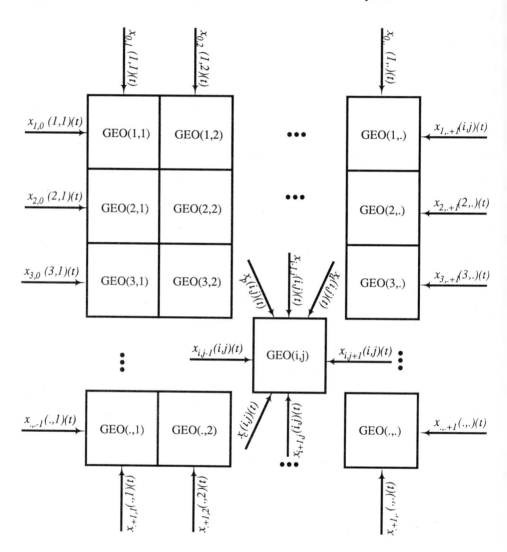

Fig. 3. Geophysical system structure showing the components, GEO(i,j) and all of the inputs to the components

The state of each component represents the current level of nitrate contamination. The output represents the spatial movement of the contamination as well as the current exposure level to any of the human population that may be exposed to the component. The state transition function provides the component mass flow balance. It contains information about the geological structure of the component.

2.3 Human System Model

The human system model contains information about the time-space population distribution and the nitrate concentration of the water supply that constitutes the exposure of the population. The model is assumed to have a structure similar to that of the geophysical system model described above. The spatial structure is represented by a matrix of system components. A one-to-one relationship is assumed to exist between the components of the human and geophysical systems. Hence, for each component of the geophysical system model, there is a corresponding component of the human system model.

Let the mathematical system model of the (i,j)th human component system be denoted

$$HUM(i,j) = (X(i,j), S(i,j), F(i,j), Z(i,j), G(i,j), T(i,j)) \qquad [4]$$

The time scale of the system will be again assumed to be discrete with each time unit equal to one day. Hence,

$$T(i,j) = \{0,1,2,...\}$$

The inputs to each component of the human system include exposure to nitrates from the water supply as well as input from the control system. The input from the control system allows the human population to receive treated water and a redistribution of the population over the spatial area. Both treatment and redistribution are possible control strategies.

The state of the human component contains information about the number of individuals who receive their water from each geophysical region, a time average of nitrate exposure dose, and the risk of an individual having cancer. The risk of an individual having cancer is based on the time average of the individual exposure dose. The state transition function calculates the population distribution assignments, the average individual exposure dose, and the individual risk that relates exposure dose and risk. The exposure-dose-risk relationship may be modeled as an instantaneous dose-response relationship.

2.4 Control System Model

The control system is the object of the design effort; hence the specifics of each design alternative must be developed for each design. Here the set of all possible inputs and all possible outputs will be established, and the particulars of the possible designs will be formulated as design alternatives. Each design alternative may not utilize every available input or may not produce each possible output. Three design alternatives are presented here.

There are many possible control system designs. Each design will consist primarily of technological implementation, such as water treatment, and/or regulation policies, such as zoning regulation of sewage treatment facilities. It is not necessary that a design consist of only one or the other but may consist of a balance of both. Examples of regulation designs include controls on fertilizer use (Pereira, 1990), source control (Samsunlu, 1990), land-use change (Chilton, 1990), and control of septic-tank effluent (Rock, 1990) to name a few. There are many methods for water treatment (Miguel, 1990; Rogalla, 1990) including biological denitrification (Cook, 1990) and ion-exchange techniques (Turkman, 1990).

Another control design that becomes apparent using the geophysical and human system models presented here is the allocation of the population to different geophysical regions for water distribution. Typically this type of a design depends on the availability of wells or surface water supplies. It also depends on the ability to route water over a distribution network without contamination. Most large metropolitan areas have distribution networks that can accomplish this type of control.

Each control system design alternative will have the ability to utilize modern principles of monitoring and analysis of the water supply (Nachtnebel, 1990). From a design point of view it will be assumed that the control system knows the state, input, output, and dynamic structure of each of the component systems. This information can be used for the particular control system design, but each bit of information will have an associated cost. This cost may be associated with the measurement or the observation of the information. For example, the concentration of nitrates in a single component of the geophysical system may be measured for a small cost if a well exists or for a large cost if a test well must be built in order to get a sample.

The control system receives inputs in the form of measurements or observations from both the geophysical system and the human system. The outputs of the control system are the effects that the control can have on the other system components.

Without considering a particular control system design, it is not possible to determine explicitly the state of the control system. It is possible to say that the state must contain at least partial information about the inputs, outputs and states of the geophysical and human system components. Each particular control system design alternative must be defined in terms of the state, the next state function and the readout function that implements a particular system design concept.

2.5 System Coupling

The relationship between system models is established through system coupling. When system models are coupled there are two factors of significant importance: component system time scale and input-output compatibility. Different component systems can operate at different time scales. For example, the geophysical system distributes nitrate through either diffusion or convection. The time constants of this system could be measured in days, whereas the effects of ingested nitrates by a human may require several days for noncancer effects and many years for cancer effect. When system components have different time scales it is important that the input-output coupling relationship account for this variation. One approach is to define equivalent time scales by scaling the system dynamics to operate at the time scale of the fastest component. Another approach is to ensure that the output is properly converted to units that are acceptable as inputs to the appropriate system. If an output of one system component is the input to another component, or is used as feedback, then it is required that the input and output sets be equal. Only under this condition will the coupling be input-output compatible.

The mathematical relationship representing the coupling between the system components is established through a system coupling recipe (Wymore, 1990). A system coupling recipe, *SCR*, is a pair denoted as

$$SCR = (\mathbf{VSCR}, CSCR) \qquad [5]$$

where **VSCR** is a vector of the component systems that are coupled. Here,

contamination system. By homomorphic model of the "real world" system, it is meant that the models presented above and the "real world" system exhibit the same behavior, except that the models cannot exhibit every possible mode of behavior of the "real world" but only those that are of interest to the control system design problem.

On the basis of these models, measures of performance are defined that characterize system behavior. These measures of performance are used to compare the validity of the mathematical models, to determine the sensitivity of the system to parameter variations, and to compare different control alternatives. Here we present a mathematical structure in which alternative designs can be compared.

It is desired to find the "best" or "optimal" control system design for the nitrate contamination system. Generally, the concept of "best" is a multicriterion decision problem, and should be called the "best compromise" or "satisfactum" (Goicoechea et al., 1982). It is desired to achieve the best system performance for the least cost. Usually these two objectives are competing and a tradeoff between performance and cost is necessary. The approach taken here is to identify performance measures that characterize system behavior in two distinct categories: input/output performance and technology performance. Input/output performance characterizes the effectiveness of a control system in terms of the risk to the exposed population. Technology performance characterizes the cost of a particular technology in implementing a control/management system design. These performance measures are then aggregated into figures of merit, or "super" performance measures, that can be used to compare systems in terms of both input/output performance and technology performance. Tradeoffs between input/output performance and technology performance can be made on the basis of the respective figures of merit.

In addition to the desire to design the "best compromise" control system, it is usually necessary for the design to satisfy other system requirements. These requirements may include the length of the system life cycle (often referred to as the planning horizon), restrictions on the available technology, and other governmental regulations. These requirements may also be taken as further objectives, and a multiple-objective decision-making scheme can be applied (Szidarovszky et al., 1976; Tecle and Duckstein, 1990).

3.1 Input/Output Performance Indices

In general, performance indices are measures, or statistics, that can be observed in both the system model and the "real world" system. Input/output performance indices can be used to answer the question: "How well does a design perform its control function?" Performance indices are computed for a single realization of the dynamical behavior of the system under consideration. Recall that it was assumed that the input process may contain uncertainties due to rainfall/runoff and other unpredictable factors; hence, the performance indices are used to characterize the system performance for a single input process realization. To highlight this fact, the k-th performance indices will be denoted $PIk(x(t))$ where $x(t)$ is the input realization. For the nitrate contamination problem, the input process $x(t)$ is the nitrate contamination that is an input to the geophysical system. Examples of input/output performance indices include:

PI(1): Expected individual risk, taken over both time and space

PI(2): Expected population risk, taken over time and space

PI(3): Average exposure dose

PI(4): Maximum exposure dose

PI(5): Minimum exposure dose

PI(6): Number of concentration measurement made on an existing well or water point source during the system life cycle

PI(7): Number of concentration measurement made on a nonexisting well or water point source during the system life cycle

PI(8): Efficiency of the treatment process

These and other performance indices can be used to compare alternative system designs on the basis of input/output performance. Each of these performance indices is compiled as an expectation over time and space.

In general, it is desirable to determine which designs perform the function of controlling the exposure to nitrates well and which perform poorly. As part of the system design process it is important to look at the values of the individual performance indices to determine the relative importance and sensitivity of the decision methodology. This inspection also provides insight into the decision maker's preference structure.

3.2 Technology Performance Indices

Technology performance measures are used to determine the utilization of technological resources such as capital and operating costs. These performance indices are defined in terms of the technology that is used to implement a control system design. Possible technologies include biological denitrification and ion-exchange techniques of water treatment. Examples of the technology performance indices include:

PI(9):   Acquisition cost of a control system design.

PI(10): Cost of measuring nitrate concentration in a sample of water.

PI(11): Cost of new wells or water supplies.

PI(12): Cost of enforcing regulation policies.

PI(13): Cost per liter to remove nitrates from the drinking water.

PI(14): Cost per liter to import less contaminated water.

PI(15): Economic impact of a control system design. For example, if a sewage treatment facility must be shut down because of high nitrate contamination, then there will be a cost associated with sewage disposal as well as an increase in unemployment.

These technology performance measures, often referred to as utilization of resource measures, represent only a few of the many possible factors that are important in considering a specific system design alternative. Some of these measures will remain constant over a single realization of the system life cycle, such as acquisition cost, and others will depend on the particular realization, such as the cost of new wells and enforcement costs. It is important to consider measures that can characterize many different design alternatives.

3.3 Figures of Merit

Figures of merit are "super criteria" and are defined as combinations of performance indices or as expectations over the set of all possible uncertainties. In water-resource applications, performance indices may be combined to obtain figures of merit as developed in Duckstein et al. (1987). Other techniques for defining figures of merit in terms of performance indices include multiattribute value theory, multiattribute utility theory (Szidarovszky et al., 1986), joint fuzzy-set membership functions (Bogardi et al., 1983), and composite programming (Nachtnebel et al., 1986).

In the previous discussion it was assumed that the uncertainties existed solely in the input process, namely, contamination. Hence, the figures of merit may be defined as expectations of functions of performance indices; this expectation would be taken over the distribution of input processes.

If the input process is denoted $X(t)$ and has known distribution F, then a figure of merit can be defined as

$$FM_k = \int Plk \; dF \qquad [14]$$

Note that the input contamination has been defined as nitrate from both point and distributed sources to each component of the geophysical system model; hence, $X(t)$ is a vector stochastic process.

Figures of merit also contain information that reflects the decision maker's preference structure. This preference structure may be reflected in terms of utility, risk or value functions. As in the definition of performance indices, figures of merit are defined for each input/output and technology element.

3.3.1 Input/Output Figure of Merit

Input/output figures of merit are criteria that reflect how well the system design performs its specified task. In addition to the input/output figures of merit which were defined directly as expected values of the performance indices by Equation [14], an additional figure of merit, engineering risk, is of primary interest. The qualification "engineering" is intended to limit the scope of the word "risk" which could encompass concepts of purely social and psychological sciences. Engineering risk has been previously defined (Bogardi et al., 1989) as the probability of a consequence (cancer). Duckstein et al. (1987) define engineering risk as the joint probability distribution function of the set of all possible values of the figures of merit. Here engineering risk will be defined as a function denoted RI and for each control system design alternative, denoted CON_i defined as:

$$RI \, (CON_1) \; = \; R \, (Hazard, \, Event, \, P(Event), \, Consequence, \, Perception) \qquad [15]$$

$$= \; R \, (\, (FM_{3,4}, \, FM_5, \, Cancer, \, P(Cancer) \, L \, (FM_3, \, FM_4, \, FM_5), \, U(L(\cdot)) \,)$$

where the hazard is the exposure to nitrates that is characterized by the expected exposure

The design of large-scale complex systems such as the one considered here is an inherently difficult problem. The approach taken in this paper has been to identify each of the important components of the system design process, from models of related system components to the identification measures of performance that reflect the important issues to be considered in the design. While the discussion presented here is far from a complete treatment of every important issue in the design, it does present a structure for understanding the design process. In addition, the relationship between the important issues of contamination, exposure and control of the nitrate pollution problem are presented in a framework that identifies the important issues for future work in this area and provides a common language among specialists discussing the system components.

Acknowledgements

Research leading to this paper has been partially supported by grants from the U.S. National Science Foundation ECS-8802350, ECS-8802920.

References

Bogardi I, Bardossy A, Curtis B (1989) Risk analysis for nitrate contaminated groundwater supplies, Working Paper, Civil Engineering Department, University of Nebraska-Lincoln, Lincoln, NE

Bogardi I, Duckstein L, Bardossy A (1983) Regional management of an aquifer under fuzzy environmental objectives, Water Resource Res 19(6):1994-1402

Duckstein L (1982) Systems approach to groundwater resources, Proceedings, International Conference on Modern Approaches to Groundwater Resource Management, vol II. Capri, Italy, October 25-27, p 13

Duckstein L, Bardossy A, Barry T, Bogardi I (1990) Health risk assessment under uncertainty: a fuzzy risk methodology. In: Haimes YY, Stakhiv EZ (eds) Risk-based decision-making in water resources, Vol IV. ASCE, NY, p 205

Duckstein L, Plate E, Benedini M (1987) Water reliability and risk: a systems framework. In: Duckstein L, Plate E (eds) engineering and reliability and risk in water resources NATO ASI Series, p 1

Goicoechea A, Hansen D, Duckstein L (1982) Multiobjective decision analysis with engineering and business applications. John Wiley, New York

Hartman PE (1982) Nitrate load in the upper gastrointestinal tract - past, present, and future. Branbury Report 12: Nitrosamines and Human Cancer, p 415

Nachtnebel HP, Hanisch P, Duckstein L (1986) Multicriterion analysis of small hydropower plants under fuzzy objectives. Annals of Regional Science 20(3):86-100 November

Szidarovszky F, Gershon M, Duckstein L (1986) Techniques for multiobjective decision making in systems. Elsevier, Amsterdam

Tecle A, Duckstein L (1990) A procedure for selecting MCDM techniques in forest resource management. Proceedings, IX-International MCDM Conference, Aug. 5-8, Fairfax, VA

Wymore AW (1990) A mathematical theory of system design SANDS, Tucson, AZ
Wymore AW (1976) Systems engineering methodology for interdisciplinary teams, Wiley Interscience, New York

NATO Advanced Research Workshop Papers

Amsden TL, US Environmental Protection Agency programs related to the agricultural use of nitrates

Bardossy A, Nitrate risk management

Boschet AF, Modification of the quality of drinking water

Cardosso SM, The nitrate content of drinking water in portugal

Chilton PJ, Alleviation of nitrate pollution of ground water by land-use change,

Cook NE, Biological denitrification of a potable water supply

Crespi M, Evidence that n-nitroso compounds contribute to the causation of certain human cancers

Dahab M, Nitrate treatment methods: an overview

Dillon PJ, Biochemistry of a plume of nitrate-contaminated ground water

Dourson M, Impact of risk-related concerns on the US Environmental Protection Agency programs

Dyhr-Nielsen M, Assessment and control of nitrogen contamination in Denmark,

Enfield CG, In-situ nitrate removal from ground water through enhanced denitrification

Fried JJ, Nitrate contamination and its control in Europe,

Forman D, Nitrate exposure and human cancer,

Ganoulis J, Nitrate contamination of surface and ground waters in Greece,

Gillham R, Nitrate contamination of ground water in Southern Ontario: A hydrogeologic perspective

Horvath G, Nitrate contamination control in the waterworks of Budapest

Hotchkiss JH, Quantifying gastric nitrate and nitrate-reductase activity in the normal and achlorhydric stomach

Kelly WE, Nitrate ground-water modeling for agriculture and other nonpoint sources

Leach SA, Some factors affecting n-nitroso compound formation from ingested nitrate in the stomach: achlorhydria, bacterial nitrosation and its modulation

Mackerness C, An in vitro chemostat model of drinking water supply systems to assess the etabolic activity of biofilm bacteria

Miguel AF, Newly developed process for removal from drinking water

Mirvish S, Factors that affect the formation of n-nitroso compounds in the body

Moller H, Epidemiological studies related to the formation of n-nitroso compounds

Mull R, Nitrate balances in aquifers

Nachtnebel HP, Principles of monitoring and analysis

Onstad C, Nitrogen management research in the US President's water quality initiative

Packer PJ, Development of a reliable measure for determining nitrate exposure for use in epidemiological studies

Pereira LS, Fertilizing and nitrate pollution

Plate E, Risk analysis for water supply from a river polluted by nitrate runoff

Rock CA, Elimination of ground-water contamination by septic tank effluent

Rijtema P, Ground-water nitrate models

Rogalla F, Experience with nitrate treatment methods

Samsunlu A, Control of nitrate sources and principles of treatment

Schepers J, Integrated water and nitrogen management practices to reduce nitrate leaching

Nitrate in drinking water was first associated with infantile methemoglobinemia by Comly (1945). The toxicity of nitrate is due to its *in vivo* reduction to nitrite and the subsequent formation of methemoglobin. Methemoglobin is incapable of binding oxygen, resulting in reduced oxygen transport from the lungs to the tissues. In normal healthy individuals, the methemoglobin reductase system, present in the erythrocytes, maintains methemoglobin levels at about 0.5 to 2.0% of the total hemoglobin (NAS, 1981). At levels above 10%, oxygen deprivation produces a bluish color in the skin and lips, and methemoglobinemia becomes clinically detectable. Weakness, tachypnea, and a rapid pulse rate are observed at levels above 25%, and death may occur at levels above 50 to 60% (NAS, 1981).

Reported cases of methemoglobin have generally occurred in infants under three months of age who have consumed water containing elevated levels of nitrates (Donahoe, 1949; Walton, 1951; Bosch et al., 1950). There are several physiological and biochemical reasons for the susceptibility of infants:

1. Exposure can be greater for infants because the ratio of total fluid intake to body weight is approximately three times that of adults.

2. Infants have an incompletely developed capacity to secrete gastric acid, and as a result, gastric pH may become high enough (pH 5 to 7) to allow nitrate-reducing bacteria to reside in the upper gastrointestinal tract.

3. The predominant form of hemoglobin present at birth, hemoglobin F (fetal hemoglobin), is more susceptible to methemoglobin formation than the adult form of hemoglobin A.

4. The enzyme NADH-dependent methemoglobin reductase, which is responsible for the normal reduction of methemoglobin (Winton et al., 1971), is less active in infants.

In 1985 the EPA established 1.0 mg nitrate-nitrogen/kg/day as the reference dose for nitrate; that is, the lifetime exposure that is likely to be without an appreciable risk of deleterious effects. The decision was based on the foregoing information and on data that suggested that infantile methemoglobinemia was the most critical effect of nitrate toxicity at low levels of nitrate exposure. This RfD was used to support the existing nitrate standard of 10 mg NO_3-N/L of drinking water.

Since 1985, however, concern has developed for the potential effects of continuous low-level exposures to nitrate, which could reduce the amount of oxygen delivered to various organs and possibly contribute to developmental, neurological, or other chronic systemic effects in children or adults. Epidemiological studies by Dorsch et al. (1984) and Arbuckle et al. (1988), for example, suggested an association between nitrates in drinking water and human birth defects. These concerns prompted the EPA to reexamine the reference dose for nitrate, and in turn, the drinking water standard of 10 mg NO_3-N/L.

2 Terminology

Definitions used throughout this paper are consistent with EPA usage (1990a). These definitions are meant for illustration only; other terms are used in different organizations and countries. These definitions include:

Critical effect. The first adverse effect, or its known precursor, that occurs as the dose rate increases.

Lowest-observed-adverse-effect level (LOAEL). The lowest exposure level at which there are statistically or biologically significant increases in frequency or severity of adverse effects between the exposed population and its appropriate control group.

Modifying factor (MF). An uncertainty factor which is greater than zero and less than or equal to 10; the magnitude of the MF depends upon the professional assessment of scientific uncertainties of the study and data base not explicitly treated with the standard uncertainty factors.

No-observed-adverse-effect level (NOAEL). An exposure level at which there are no statistically or biologically significant increases in the frequency or severity of adverse effects between the exposed population and its appropriate control; some effects may be produced at this level, but they are considered neither adverse nor precursors to specific adverse effects. In an experiment with several NOAELs, the regulatory focus is primarily on the highest one, leading to the common usage of the term NOAEL as the highest exposure without adverse effect.

Reference dose (RfD). An estimate (with uncertainty spanning perhaps an order of magnitude) of a daily exposure to the human population (including sensitive subgroups) that is likely to be without appreciable risk of deleterious effects during a lifetime. RfD = [NOAEL or LOAEL ÷ (UF · MF)].

Uncertainty factor (UF). One of several, generally 10-fold factors, used in operationally deriving the reference dose (RfD) from experimental data. Uncertainty factors are intended to account for (1) the variation in sensitivity within the human population, (2) the uncertainty in extrapolating animal data to humans, (3) the uncertainty in extrapolating from data obtained in a study that is of less-than-lifetime exposure, (4) the uncertainty in using LOAEL data rather than NOAEL data, and (5) the inability of any single study to adequately address all possible adverse outcomes in man.

3 Review of Human Health Effects

3.1 Methemoglobinemia

The Reference Dose (RfD) for nitrate is based on a composite of human clinical and epidemiologic studies in which the critical effect associated with ingestion of nitrate-containing water, clinical methemoglobinemia (i.e., visible signs of methemoglobinemia such as bluish skin color), was observed to occur only in infants aged zero to six months. No cases of nitrate-induced clinical methemoglobinemia have been documented for older infants and for children ranging in age from one to eight years (Craun et al., 1981). The susceptibility of young infants to nitrate-induced methemoglobinemia can be attributed to the developmental immaturity of the digestive and hematopoietic systems at this early age. This immaturity includes the following characteristics: (1) a reduced capacity to secrete gastric acid, which may lead to an increase in the pH of the upper gastrointestinal tract to a level of 5 to 7, leading to bacterial proliferation and nitrate catalysis to nitrite; (2) a deficiency in methemoglobin reductase or its cofactor, NADH; and (3) an apparently greater susceptibility of fetal than adult hemoglobin to oxidize to methemoglobin (Fan et al., 1987).

Since nitrate-induced methemoglobinemia has not been clearly demonstrated in experimental animal models, such data do not appear to be relevant to the assessment of hazard in humans. Thus, nitrate belongs to an emerging class of compounds, including arsenic, barium, manganese and nickel, in which the critical effect associated with compound exposure is unique to humans. In these cases, a "weight-of-evidence" approach, combining the results of several human studies, is appropriate. Several of these studies are discussed below and summarized in Table 1.

Table 1. Nitrate toxicity studies on methemoglobin and associated clinical signs in infants

Study	Number	Concentration (mg/L) Nitrate as Nitrogen	Effect
Bosch et al., 1950	2 cases	10-20	Questionable methemoglobin with clinical signs
	25 cases	21-50	Methemoglobin with clinical signs
	53 cases	51-100	Methemoglobin with clinical signs
	49 cases	>100	Methemoglobin with clinical signs
Cornblath and Hartman, 1948	4	70^b	Methemoglobin of 5.3%
	4	140^b	Methemoglobin of 7.5%
	$?^c$	140^b	Methemoglobin of 11% and early signs of cyanosis
Donahoe, 1949	2 cases	11 or 40 with bacterial contamination	Intense cyanosis
Ewing and Mayon-White, 1951	2	46 or 22 with bacterial contamination	Death or transfusion
Jones et al., 1973	1	25	Severe cyanosis
Simon et al., 1964	89	0	None
	38	11-23	None
	25	>23	Statistically increased methemoglobin
Toussaint and Selenka, 1970	34	34	Methemoglobin of 3%
Walton, 1950	0	0-10	No methemoglobin with clinical signs
	5	11-20	Methemoglobin with clinical signs
Total cases for		21-30	Methemoglobin with clinical signs
which data are	36	31-50	
available =	81	51-100	Methemoglobin with clinical signs
214	92	>100	Methemoglobin with clinical signs
Winton et al., 1971	63	4^b	None
	23	$1-7^b$	None
	20	$7-14^b$	None
	5	$14-23^b$	Higher than normal methemoglobin

[a]   With some evidence of bacterial contamination in 88% of water samples
[b]   Estimated assuming 0.64 L/day for a 4 kg infant
[c]   Unspecified number of infants with history of previous cyanosis due to nitrate in well water contaminated with nonpathogenic bacteria

3.2 Case Studies

Bosch et al. (1950) evaluated 139 cases of clinical methemoglobinemia reported by physicians in Minnesota in which 14 deaths occurred. The youngest case was eight days old and the oldest was five months. Ninety percent of the cases occurred in infants under two months of age. Nitrate-nitrogen concentrations of well water used to prepare drinking formula for the infants were analyzed for 129 cases. None of the wells contained less than 10 mg/L nitrate-nitrogen. Two wells (1.5%) contained 10 to 20 mg/L; ingestion of water from these wells was associated with questionable diagnosis of methemoglobinemia. Twenty-five wells (19%) had concentrations of 21 to 50 mg/L; 54 (41%) of 51 to 100 mg/L; and 49 (38%) of greater than 100 mg/L. Coliform organisms were detected in 45 of 51 samples (88%) tested for bacterial contamination, indicating the likely presence of nitrite contamination as well.

A total of 278 cases of infant clinical methemoglobinemia, involving 39 deaths, in 14 states were reported in a 1950 public health survey (Walton, 1951; Bosch et al., 1950). Of 214 cases for which data on nitrate-nitrogen levels were recorded, none occurred in infants consuming water containing less than 10 mg/L. Five cases (2%) were reported in infants exposed to 11 to 20 mg nitrate-nitrogen; 36 cases (17%) in infants exposed to 21 to 50 mg/L; and 173 (81%) in infants exposed to more than 50 mg/L. A single case study, involving a four-week old infant, showed methemoglobinemia with clinical signs at a drinking water level of approximately 17 mg NO_3-N/L (Vigil et al., 1965).

3.3 Clinical Studies

Cornblath and Hartmann (1948) supplied nitrate-containing water in formula to four infants ranging in age from less than two weeks to eleven months at doses of 11 mg nitrate-nitrogen/kg/day for 2 to 18 days. (Assuming an average daily infant consumption of 0.16 L/kg, the administered dose corresponds to approximately 70 mg/L.) The highest measured methemoglobin concentration was 5.3%. Neither symptoms of cyanosis nor other clinical signs of methemoglobinemia were observed. Four infants aged two days to six months were subsequently given a doubled dose corresponding to approximately 140 mg/L nitrate-nitrogen. The highest concentration of measured methemoglobin was 7.3%; no clinical symptoms were detected.

Simon et al. (1964) measured methemoglobin levels in 89 healthy infants who received nitrate-free water, 38 infants who received water containing 11 to 23 mg nitrate-nitrogen/L and 25 infants receiving water containing more than 23 mg nitrate-nitrogen/L. The mean methemoglobin levels in these groups were 1.0, 1.3 and 2.0%, respectively. No clinical signs of methemoglobinemia were observed.

Toussaint and Selenka (1970) administered formula prepared with water containing 34.5 mg/L nitrate-nitrogen to 34 healthy infants. Average methemoglobin levels rose to 2 to 3% within 2 days and remained elevated for up to 10 days. No clinical signs of methemoglobinemia were reported.

Gruener and Toeplitz (1975) administered formula prepared with water containing approximately 25 mg of nitrate-nitrogen/L to 104 infants ranging from one week to ten months of age for three days. Average methemoglobin levels rose to 1.3%. No clinical signs of methemoglobin were reported.

3.4 Epidemiologic Studies

Winton et al. (1971) measured methemoglobinemia levels associated with ingestion of nitrate in 111 infants aged less than two weeks to six months. None of the 106 infants ingesting 0.6 to 2.3 mg/kg/day nitrate-nitrogen (equivalent to approximately 4 to 14 mg/L) had methemoglobin levels outside the normal range. Of five infants receiving 2.3 to 3.6 mg/kg/day nitrate-nitrogen (equivalent to 14 to 23 mg/L), three had methemoglobin levels somewhat higher than average, but none exceeded 5.3%. No signs of clinical methemoglobinemia were detected.

Craun et al. (1981) examined the effect of ingestion of nitrate-containing drinking water in 102 children aged from one to eight years. Concentrations of 22 to 111 mg/L of nitrate-nitrogen were not associated with increased methemoglobin concentrations. None of the children studied had levels of methemoglobin that could be considered high or above normal.

Although clear-cut diagnosis of clinical methemoglobinemia and accurate measurements of nitrate concentrations in drinking water are lacking in some of the studies, one consistent

3. Uncertainties in exposure

In this paper the physical system (ground-water transport system) is assumed to be identified. The other uncertainties and their influence on the final risk management are considered. A numerical example is used to illustrate the different steps.

Uncertainty is handled in a fuzzy framework. One reason for doing this is that fuzzy sets offer a combination rule (the min-max combination) which is rather simple and transparent. In fuzzy sets there is no need to define complicated dependencies between different variables which may be difficult to specify. These often subjective dependence/independence assumptions have a major influence on the results in the case of probabilistic methods. Further details on the approach of the techniques presented and used in this paper can be found in Bogárdi et al., 1990; Bárdossy et al., 1990a; Duckstein et al., 1990.

2 Risk Assessment

The two main elements of nitrate risk characterization are the assessment of exposure dose and that of dose-response. The uncertainties present in both elements contribute to the overall uncertainty of nitrate risk estimates and are briefly reviewed.

Uncertainties in exposure dose assessment stem from uncertainties in individual dose assessment and variation of individual doses in the exposed population. Exposure uncertainty has been addressed by many investigators using techniques such as probability theory (Crump and Howe, 1985; Finkel and Evans, 1987; Hornung and Meinhardt, 1987), the entropy concept (Lind and Solana, 1988), and fuzzy-set analysis (Feagans and Biller, 1980, 1981). Concerning dose-response assessment, four key points can be identified (Sielken, 1988): (1) exposure dose scale, (2) dose-response model, (3) interspecies extrapolation, and (4) data set. These points are now developed.

2.1 Exposure Dose Scale

Three possible dose scales for dose-response modeling can be distinguished:
1. Administered (or applied) dose scale
2. Delivered (or target) dose scale
3. Biologically effective dose (BED) scale

The administered dose refers to the amount of nitrate reaching an individual. The delivered dose is the amount of nitrate reaching the target organ. A pharmacokinetic model can describe the relationship between administered dose and delivered doses. The net amount effective at the target site defines the biologically effective dose (BED).

2.2 Dose-response Model

Dose-response models relate the exposure dose to the probability of cancer. Following the historical development of such models, three main groups can be distinguished (Sielken, 1988; Zeise et al., 1987):

1. Time-invariant, mostly statistically based models (one-hit, multi-hit, multistage; and probit, logit, Weibull). The first three models appear to yield higher estimates in the environmental dose domain than the last three models.
2. Time-varying, mostly statistically based models (multistage-Weibull, Hartley-Sielken).
3. Biologically based cancer (cell-growth) models.

The most important uncertainty issue is the model-choice problem, since often there is no biological basis to prefer one statistically based model to another. Test data allow the use of several statistical models which lead to very different risk estimates. In general, biologically based models are preferred to statistically based ones. Cohen and Ellwein (1988) claim that biologically based models are independent of the specific chemical under consideration. However, the input for such models, or BED, is often unavailable, and the models are quite uncertain. In such situations a logical approach seems to be to consider several possible dose-response models and use expert judgment to assess the "degree of correctness" of each possible model.

2.3 Interspecies Extrapolation

Although interspecies extrapolation is controversial, it is necessary in almost every case. Even when human data are available, animal studies with controlled exposure doses should be utilized to update cancer responses estimated from human data. Recent findings support the general use of animal data to evaluate carcinogenic potentials in humans (Allen et al., 1988 and discussions following the paper). However, it is necessary to emphasize that interspecies extrapolation is uncertain since, for example, it may not account for competing risks, the existence of a no-effect threshold, or the recently discovered fact that animals react differently to nitrates than humans.

In general there are three possibilities for conversion: (1) body surface area correction, (2) body-weight correction, and (3) no correction. Often there is not enough information available to select one "true" interspecies correction factor. However, as in the case of dose-response models, a "degree of correctness" characterization is possible (Sielken, 1988). This degree of correctness is related to the risk of assessment. For instance, being dispensed of using interspecies correction when BED is based on a biological cancer model yields a high degree of correctness. The degree of correctness is smaller when the administered dose is used without an interspecies correction factor.

2.4 Data Set (Dose Response)

The larger the data set used to estimate cancer risk, the more certain is the estimate. However, there are "good" data and "fair" data, and direct (human) and indirect (animal bioassay) data, observed under different experimental conditions. Which data set should be considered? Often, one single data set is used to estimate nitrate risk; however, another data set may result in a quite different risk. Allen et al. (1988) recognize 11 groups of decision parameters involved in animal data. All relevant information should be used to estimate nitrate risk. The problem is to decide how to combine data sets which are of varying quality.

3 Decision Problem Formulation

3.1 The Event/Decision or Probability Tree

An event/decision-tree framework will be used to assess cancer risk assessment. As explained, for example, in Benjamin and Cornell (1970) or Ang and Tang (1984), uncertain phenomena and decisions can be represented in the form of event trees for descriptive purposes and decision trees for decision-making purposes. In this investigation trees will involve problems of cost-effectiveness, resource allocation and risk management. Also, several alternative approaches for analyzing uncertainty along a given tree pathway may be used, such as the mixed probability/fuzzy-set approach used in Duckstein et al. (1990). Accordingly, the term possibility tree will designate either an event tree or a decision tree, or a combination of both types of trees.

A decision tree for nitrate is sketched in Fig. 1. Nitrate intake appears to be a major contributor of gastric nitrite, which produces nitrosoamines and nitrosoamides, which in turn are etiologic agents for human gastric cancer. Fig. 1 shows the three pathways that have been selected, leading to three different risk estimates. These pathways are as follows:

1. Uncertain nitrate exposure
2. Use of delivered dose
3. One-hit, two-stage and logit models for dose-response relationship
4. No interspecies correction
5. Use of the data in Terracini et al. (1967), also shown in Table 1

Table 1. Liver tumors in rats fed by N-nitrosodimethylamine (NDMA) (after Terracini et al., 1967)

Dose in diet (ppm)	No. of animals with liver tumor	No. of animals on test	Credibility level
0	0	41	0.9
2	1	37	0.8
5	8	83	1.0
10	2	5	0.4
20	15	23	0.7
50	10	12	0.6

4 Decision Analysis by Fuzzy Logic

4.1 Interval Analysis and Fuzzy Sets

The simplest method of considering uncertainty is to perform an interval analysis. An uncertain parameter of nitrate risk assessment, such as interspecies correction, takes on any value within such an interval. With more information on the uncertain parameter the interval can be "shrunk"; that is, we can determine that the parameter can take only certain value(s) within the interval (see Fig. 2). Fuzzy-set theory, which can be used with very few and weak prerequisite assumptions, is selected as an extension of interval analysis.

4.2 Fuzzy Sets, Fuzzy Numbers, Extension Principle

The basic definitions of fuzzy sets, fuzzy numbers, and fuzzy operations are summarized

The equivalence of the two formulations can be proved using the extension principle (Bárdossy, 1990). In fuzzy regression, contrarily to probabilistic regression, the approximate equality from Equation [2] has to be fulfilled for each t which defines a set of constraints as shown below. Depending on the form of the function f, these constraints may be simplified in various ways.

The "goodness" of a fuzzy regression is measured by means of so-called vagueness criteria. The vagueness V of a fuzzy regression can be defined in several different ways. The different vagueness criteria refer to the parameter vector **a** or to the fuzzy function f. Further mathematical details on fuzzy regression can be found in Bárdossy et al. (1990b).

Generally speaking, the minimization of vagueness leads to a mathematical programming problem with linear constraints. In linear regression, a linear programming formulation follows as described in Bárdossy (1990). The nonlinear case and further mathematical details on fuzzy regression may be found in Bárdossy et al. (1990b).

Table 2 gives an example of results obtained by fuzzy regression for the case of a two-stage model, using the data of Table 1. Fig. 3 shows the graph of the fuzzy dose-response curve.

Table 2. Regression results for the two-stage model using actual prediction vagueness

Dose (ppm)	Probability of Tumor Development (%)		
	$\mu=1.0$	$\mu=0.5$ (R) (upper)	$\mu=0.5$ (L) (lower)
0.01	0.0270	0.0390	0.0151
0.02	0.0541	0.0780	0.0302
0.05	0.135	0.195	0.0755
0.10	0.270	0.390	0.151
0.25	0.677	0.973	0.379
0.50	1.36	1.94	0.764
1.00	2.72	3.87	1.55
2.00	5.46	7.69	3.17
5.00	13.75	18.76	8.44
10.00	27.49	35.66	18.27
25.00	63.0	72.6	50.1
50.00	92.8	96.0	86.9

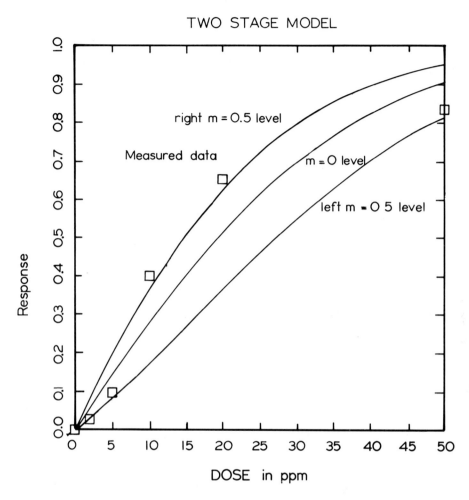

Fig. 3. Dose-response relationship for N-nitrosodimethylamine obtained with fuzzy regression (two-stage model)

4.4 Risk Estimates

Cancer risk has been estimated using three different human nitrate doses using the fuzzy dose-response models obtained via fuzzy regression. Three different doses were considered:

1. Fixed, 20 mg/day;

2. Random dose with lognormal distribution, mean = 20 mg/day, standard deviation = 7 mg/day;

3. Fuzzy dose, approximately 20 mg/day (Fig. 4)

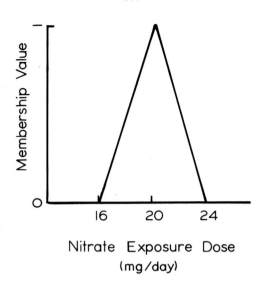

Fig. 4. Nitrate exposure as a fuzzy number "about 20"

Table 3 gives the fuzzy risk estimates, which differ substantially, for the one-hit, two-stage and logit models. Fig. 5 explains the steps already completed for fuzzy nitrate risk assessment. The next step consists in combinating these different fuzzy risk estimates so as to obtain a single fuzzy number.

Table 3. Cancer risk estimates: R1 (one-hit), R2 (two-stage), R3 (logit model)

Human Nitrate Dose mg/day	R1 Membership Value			R2 Membership Value			R3 Membership Value		
	0.5 left	1.0	0.5 right	0.5 left	1.0	0.5 right	0.5 left	1.0	0.5 right
Fixed (20)	1.18	2.60	4.33	1.09	2.29	3.75	0.003	0.004	0.007
Lognormal (20,7)	1.35	2.96	4.93	1.24	2.61	4.28	0.004	0.007	0.011
Imprecise ($\approx$ 20)	0.96	2.60	5.24	0.88	2.29	4.54	0.002	0.004	0.009

4.5 Combination of Various Risk Estimates

Each pathway of the possibility tree yields a different fuzzy risk estimate. These estimates must be combined and/or ranked. Suppose that fuzzy numbers R1 ,..., membership functions mR1(t),...,mRI(t) describe the cancer risk assessment procedure, but are based on different pathways. An overall fuzzy number R has to be found; thus the fuzzy numbers, here, risks

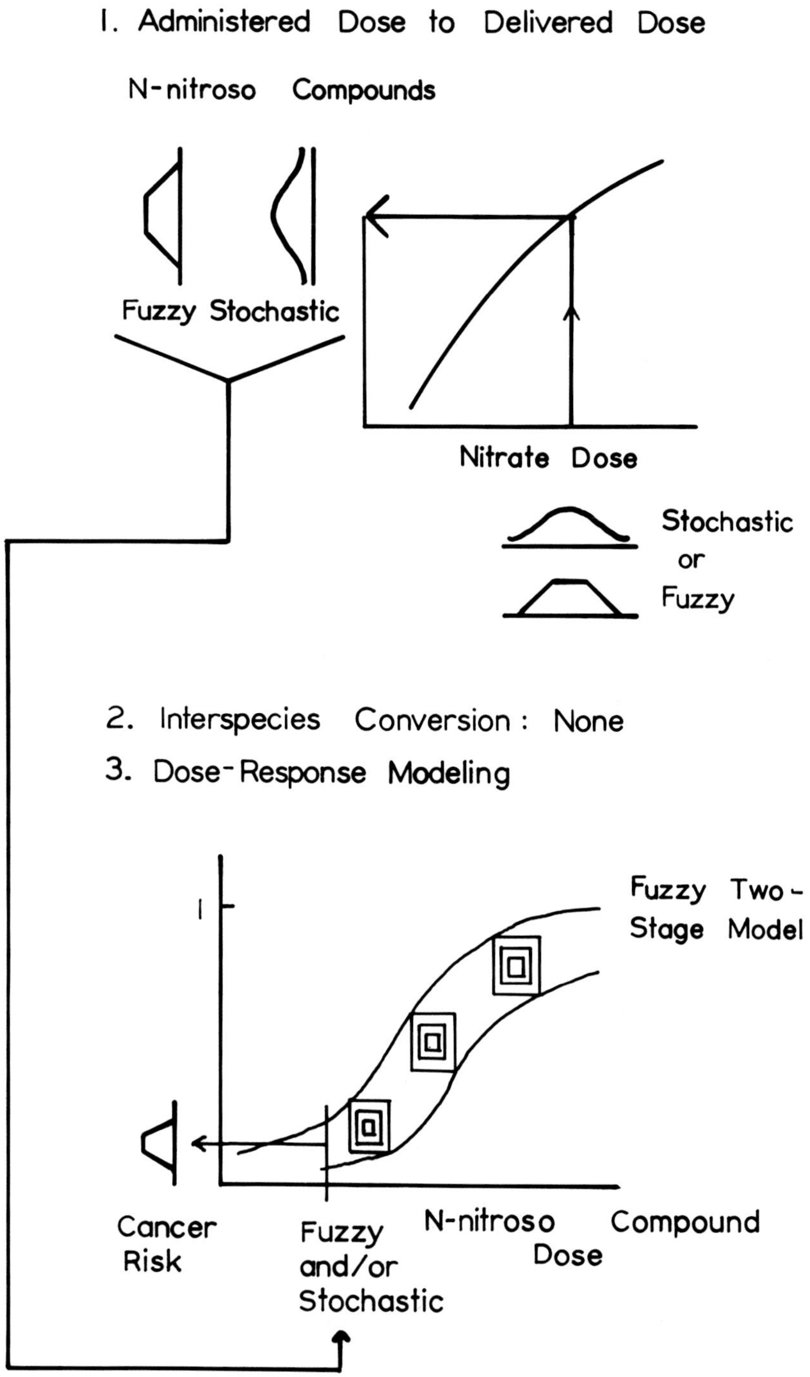

Fig. 5. Estimation of nitrate risk using possibility tree in Fig. 1

PANEL DISCUSSION: INTEGRATION, THE SYSTEMS APPROACH

Lucien Duckstein, Panel Chair and Reporter
University of Arizona
Systems and Industrial Engineering Department
Tucson, Arizona 85721 U.S.A.

Panel Members: A. Bardossy,[1] P.J. Chilton,[2] M.L. Dourson,[3] D. Forman,[4] and
D. Martin[5]

The following four questions were addressed:

1. Do we have a common definition of the nitrate contamination problem? If not, can we agree on one?

2. What are the gaps in our knowledge of the effects of nitrate on the hydrologic cycle and on human-related components of the system?

3. What are the trade-offs between controlling NO_3^-, NO_2^-, and other substances, and the corresponding costs or regrets?

4. How is risk analysis performed under the various kinds of uncertainties described during the ARW?

A summary of the panel and audience responses to these four questions follows.

1 Common Definition of the Nitrate Contamination Problem

Each professional discipline represented at the ARW appears to have a different view on the salient features of the nitrate contamination problem. For example, epidemiologists such as FORMAN consider most of the evidence of effect on human health unconvincing; government agency scientists such as DOURSON look at the risk-minimizing regulatory aspects;

[1] A. Bárdossy, Institute for Hydrology and Water Resources, University of Karlsruhe Kaiserstr. 12, D-7500 Karlsruhe, Germany.
[2] P.J. Chilton, British Geological Survey, Maclean Building, Crowmarsh Gifford, Wallingford OX10 8BB, U.K.
[3] M. Dourson, U.S. EPA, 26 W. Martin Luther King Drive, Cincinnati, OH 45268 U.S.A.
[4] D. Forman, Cancer Epidemiology Unit, Imperial Cancer Research Fund, Gibson Building Radcliffe Infirmary, Oxford OX2 6HE, UK.
[5] D. Martin, Biological Systems Engineering, University of Nebraska-Lincoln, Lincoln, NE 68583-0726 U.S.A.

NATO ASI Series, Vol. G 30
Nitrate Contamination
Edited by I. Bogárdi and R. D. Kuzelka
© Springer-Verlag Berlin Heidelberg 1991

NATO ASI Series G

NATO ASI Series G

Printing: Druckhaus Beltz, Hemsbach
Binding: Buchbinderei Schäffer, Grünstadt